# THE WONDER BOOK OF CHEMISTRY

# THE WONDER BOOK OF CHEMISTRY

JEAN HENRI FABRE

Translator

FLORENCE CONSTABLE BICKNELL

GYAN

*Published by*

Gyan Publishing House
5, Ansari Road
Daryaganj, New Delhi-110002
Phone: 011-47034999, 9811692060
E-mail: books@gyanbooks.com

*Distribution Network*
gyanbooks.com
India, USA, Canada, UK, Australia



ISBN: 978-93-7169-038-6 (HB)

First Published, 1922

2nd Impression 2025

*Printed at:* Gyan Press, Delhi.



**THE WONDER BOOK OF CHEMISTRY**
Author: JEAN HENRI FABRE

# THE WONDER BOOK OF CHEMISTRY

BY

JEAN-HENRI FABRE

Author of "The Story-Book of Science," "Our Humble Helpers," "Field, Forest and Farm," "The Secret of Every Day Things," "Animal Life in Field and Garden," etc.

TRANSLATED FROM THE FRENCH BY
FLORENCE CONSTABLE BICKNELL

NEW YORK
THE CENTURY CO.
1922

# TRANSLATOR'S PREFACE

"What is the use of a book without pictures or conversations?" asks Alice, disgustedly, just before taking her departure for Wonderland, where she finds no lack of animated discourse.

This book, like its predecessors in the series, is conversational in form and has as many pictures as the subject-matter calls for.

All boys and some girls, as well as their elders, take more or less interest in the marvels of chemistry. To give an elementary but useful knowledge of these marvels, chiefly by means of simple experiments clearly described by the writer and easily performed at home by any wide-awake young reader, is the object of the following talks by "Uncle Paul."

The personal, biographical interest of the book is not to be overlooked. The boys Jules and Emile are the author's own children, faithfully portrayed even to the names they bear. In his captivating fashion the man of vast learning makes himself at once teacher and comrade to his young hearers, and we learn that "his chemistry lessons especially had a great success. With apparatus of his own devising and of the simplest kind he could perform a host of elementary experiments, the apparatus as a rule consisting of the most ordinary materials, such as a common flask or bottle, an old mustard-pot, a tumbler, a goose-quill or a pipe-stem. A series of astonishing phenomena amazed their wondering eyes. He made them see, touch, taste, handle, and smell, and always 'the hand assisted the word,' always 'the example accompanied the precept,' for no one more fully valued the profound maxim, so neglected and misunderstood, that 'to see is to know.' " Though living creatures necessarily claimed the naturalist's first affections, he none the less "animates even the simple elementary bodies, celebrating the marvelous activities of the air, the violence of chlorin, the metamorphoses of carbon, the miraculous bridals of phosphorus, and the 'splendors which accompany the birth of a drop of water.' "

Concerning the eager young pupils, Jules and Emile, by this time well known to all readers of the series, a still further word may not be out of place. Emile, the younger, the "giddy-pate" of the narrative, impulsive and full of boyish curiosity and vigorous young life, is drawn for us with fidelity and a delightful touch of humor by the loving father. Jules is shown to us as more sedate and gifted with finer qualities, and with his grief-stricken parent we mourn his early death. "He was a youth of great promise, 'all fire, all flame'; of a serious nature; an exquisite being, of a precocious intelligence, whose rare aptitudes both for science and literature were truly extraordinary. Such too was the subtlety of his senses that by handling no matter what plant, with his eyes closed, he could recognize and define it merely by the sense of touch. This delightful companion of his father's studies had scarcely passed his fifteenth year when death removed him. A terrible void was left in his heart, which was never filled. Thirty years later the least allusion to this child, however tactful, which recalled this dear memory to his mind, would wring his heart, and his whole body would be shaken by his sobs.'' * In a memorial foreword to the second volume of his "Souvenirs Entomologiques" the bereaved father pays loving tribute to this lost son and fellow-worker.

Thus it is that the following chapters will be found to have a human and personal appeal to supplement their scientific interest. May they yield both pleasure and profit to their readers!

---

* The quoted passages are from Dr. C. V. Legros's "La Vie de J.-H. Fabre, Naturaliste," translated by Mr. Bernard Miall under the title, "Fabre, Poet of Science," and published by The Century Co.

# CONTENTS

# CONTENTS

# THE WONDER BOOK OF CHEMISTRY

# THE WONDER BOOK OF CHEMISTRY

## CHAPTER I

### INTRODUCTION

UNCLE PAUL is a man of some learning who waters his lettuce-plants and weeds his cabbages and turnips in the quiet of a humble little village. Staying with him are his nephews, Emile and Jules, young scholars already grappling with the intricacies of the rule of three and the pitfalls of the past participle, and both of them very eager to learn. Jules, the elder, is even now beginning to suspect that school will not have taught him everything when he has mastered his grammar and arithmetic. Their uncle does his best to encourage the boys' desire for knowledge, convinced as he is that in the stern battle of life our best weapon is a trained intellect.

For some time past his family had noticed in him an unusual preoccupation. There was ripening in his mind a plan for teaching his nephews the rudiments of chemistry, that science so fruitful in its practical applications.

"What are these dear children going to be, some day?" he asked himself. "Will they be manufacturers, artisans, mechanics, farm laborers, or what? Who knows? One thing, at any rate, is certain, and that is, whatever direction their activity takes it will be to their advantage to be able to give an account of the things they have accomplished. A little science is something that they must have. I should like my nephews to know what air is, and water; why we breathe, and why wood burns; the nutritive elements essential to plant life, and the constituents of the soil. And it is no vague and imperfect knowledge from hearsay I would have them gain of these fundamental truths, on which depend agriculture and the industrial arts and our health itself; I would have them know these things thoroughly from their own observation and experience. Books here are insufficient, and can serve merely as aids to scientific experiment. But how shall we manage it?"

In this wise did Uncle Paul ponder over his project, a project involving grave difficulties, such as the want of a laboratory and of all those ingenious devices without which it would at first seem impossible to undertake any serious experiments in chemistry, the only appliances at hand being the commonest of household utensils,—bottles and phials, jars and pitchers, plates and cups and earthen bowls, drinking-glasses and old mustard-pots. It is true the distance to town was not great. For special occasions, but within the very modest limits set by an imperative economy, a few drugs and glass implements might be bought. Ten francs must be made to cover these extraordinary purchases. How, then, to impart some useful knowledge of chemistry with the help of little more than such simple appliances as the village could furnish —that was the problem.

But in the end it came about that one day Uncle Paul announced to his nephews that he proposed to enliven the monotony of their regular studies by introducing a little diversion. Without using the word "chemistry," which would have meant nothing to them, he spoke of certain interesting things he had to show them, of various wonderful experiments to be performed. Lively and curious, as are all children, Emile and Jules greeted this announcement with enthusiasm.

"When shall we begin?" they asked. "To-morrow—to-day?"

"To-day, very soon. Give me five minutes for my preparations."

## CHAPTER II

### MIXING AND COMBINING

NO sooner said than done. Uncle Paul went to his neighbor the locksmith and from among the files on the artisan's work-bench selected something and wrapped it up in a piece of paper. Then he visited the apothecary and for a few cents bought a drug which he also wrapped in a bit of old newspaper, after which he returned home with his two packages.

"What is this?" he asked, opening one of the parcels before the children.

"It is a yellow powder that makes a little crackling sound when you rub it between your fingers," replied Emile. "I think it must be sulphur."

"And I," added Jules, "am sure it is sulphur. But we 'll soon see."

So saying he took a pinch of the yellow powder and dropped it on some live coals from the kitchen fire, whereupon it began to burn with the blue flame and the suffocating odor of a sulphur match.

"That proves it, I hope," cried the lad, much pleased with himself at having found a quick way to demonstrate the nature of the substance offered by his uncle. "It is sulphur and nothing else, for that is the only thing that burns with that blue flame and that smell that makes you cough."

"Yes, my boys," assented their uncle, "it is sulphur powdered very fine and called flowers of sulphur. And now what is this?"

He opened the second package and displayed its contents, consisting of a powdered metal, the fact of its being a metal showing clearly in its glittering particles.

"That looks very much like iron filings," declared the younger of the two observers.

"It does more than look like them," asserted the other; "it really is iron filings. Uncle Paul, you must have got them from the locksmith's."

"Though I must congratulate Jules on his cleverness and quickness," rejoined Uncle Paul, "I ought at the same time to warn him against jumping to conclusions. In the studies we are about to take up together it is best to exercise careful scrutiny before venturing on any assertions, as otherwise one would run the risk of making frequent mistakes. You say these metallic particles are iron filings; but lead filings, tin filings, zinc filings, iron filings—all are of very much the same appearance, being all light in color and having a bright luster. You declared the yellow powder to be sulphur after you had proved it by dropping a pinch on burning coals. Now find an equally decisive proof that these filings are of iron."

The boys put on their thinking-caps and looked at each other in mutual questioning, but no happy thought came at their bidding. To what test could they put those filings to prove that they were indeed of iron? It was a puzzling problem, that was certain. But at last Uncle Paul started the boys on the right track.

"How about the magnet," said he, "that horseshoe shaped piece of iron bought by Jules at the last fair and added to his little cabinet of apparatus for making experiments in physics? Would n't it be just the thing to help you out in your present perplexity? Many a time I 've seen you amusing yourselves with that magnet by making it draw to itself bits of iron, nails, needles. Does it have the same effect on lead?"

"No," replied Jules; "I have never been able to make it take up the least little bit of lead, though it will lift much heavier weights of iron,—a key, for example."

"Does it attract tin?"

"No; no more than lead."

"And zinc and copper—does it have any effect on them?"

"No more than on lead or tin. Ah! Now I have it. The magnet attracts only iron. That 's the test we 're after. Now we 'll see."

Thereupon Jules ran upstairs, two steps at a time, and hastened to his cupboard where, on a pine shelf, were arranged his books and his little pieces of apparatus,—simple appliances and mostly of his own make. Eagerly catching up his magnet, he ran downstairs and brought it almost in contact with the filings. Immediately there were clusters of them clinging to the two ends of the magnet, forming long beards of bristling appearance.

"See there," cried the lad, "how it makes the filings come to it! I am sure now they are iron, nothing but iron."

"Yes, my boy, they are iron," assented his uncle; "and it was the locksmith's work-bench that furnished me with the filings. Now, with this iron and this sulphur, which we have just proved to be iron and sulphur beyond any doubt, we will enter upon our study of chemistry. Give your attention to what I am about to do."

So saying, he emptied on a large sheet of paper both the flowers of sulphur and the iron filings, after which he mixed them thoroughly together by shaking the paper like a sieve and stirring its contents with his fingers.

"Look, now," said he; "what have we on the paper?"

"Oh, that 's easy enough," Jules made answer; "it 's just a mixture of sulphur and iron filings."

"Yes, a mixture; and could you still tell the one substance from the other, all mixed together as they are?"

"Nothing easier," answered Emile, examining closely what was on the paper. "Here, for instance, are some grains of sulphur; I know them by their yellow color; and here are some of the iron filings, as you can tell by their shiny look."

"And would you undertake to separate the particles of one kind from those of the other,—to sort them all out?"

"Why not, if it really had to be done? I have good eyes, and with the help of a pin I could gather all the sulphur together on one side and all the iron on the other. Only, I doubt whether my patience would hold out to the end."

"Yes, it certainly would be a rather longer job than picking over a plate of beans; and Emile's patience, however great it may be, would be hardly equal to the task. Still, the thing is not impossible. In that little heap, which has now neither the yellow color of pure sulphur nor the lustrous gray of pure iron, but which has at once something of the two colors and is consequently of a greenish appearance —in that little heap of matter, I say, an eye of sufficient patience and a hand of sufficient dexterity could, between them, see and separate what is sulphur from the iron. But there are other ways of making the separation. Who will find one? Come, now, set your wits to work."

"I have it!" cried Jules, passing the ends or poles of his magnet back and forth through the mixture.

"Just what I was going to propose," said Emile, "if Jules had given me a moment to think about it. Now that Uncle has reminded us of the magnet, the rest comes of itself."

"To hit on the way out of a difficulty after a moment's reflection is all very well, my young friend," rejoined his uncle; "but to hit on it immediately is still better. However, you will get even with Jules very soon, I am sure. Now let us see how his method of sorting the two substances succeeds."

Jules went on passing his magnet through the mixture of iron filings and sulphur, with the result that the metallic particles were attracted to the two poles of the magnet and clung to them, while the sulphur was left behind. Again and again the magnet was plunged into the heap, and each time it was withdrawn loaded at its two extremities with long and thick beards of filings which the young operator detached with his finger-tips, and placed at one side. Not a particle of the sulphur clung to the magnet, or at least not by the force of attraction, the magnet exerting no such force on sulphur; and if any scattering particles were found among the iron filings set aside by themselves, it was simply because they had become enmeshed among the grains of metal. A second, similar sorting very easily separated them.

"That 's the way to do it!" exclaimed Jules, delighted with the success of his operations. "That 's the way, see! The magnet comes out each time loaded with filings, and the sulphur is left behind. If I went on, it would n't take me more than ten minutes to separate all the iron on the paper from the sulphur."

"It is unnecessary to continue, my dear child," said Uncle Paul. "Your method is perfect, being both expeditious and unfailing in its results. Now put the iron filings back with the sulphur and mix the two well together. Your magnet, so serviceable to us in this process of sorting the two substances, is not at the disposal of every one. Would n't it be possible to get along without it, to make the desired separation in some other way? It is well, it is even indispensable, especially for us with our meager outfit, to learn how to do without what we do not possess, and nevertheless to attain results. Let us, then, dispense with our magnet and find some other way to separate the iron filings and the sulphur. Think a moment. I will help you. Which is the heavier of the two substances, the sulphur or the iron?"

"The iron," replied the two young chemists.

"And what would the iron do if we threw it into water?"

"It would sink to the bottom."

"And the sulphur—what would that do? I mean finely powdered sulphur, flowers of sulphur, not sulphur in the lump, for that too would sink in water."

"I see!" Emile made haste to answer, lest he should again be outstripped in this race of wits by his elder brother. "I see! I will throw the whole mixture into a glass of water and the iron will sink to the bottom, but the sulphur—wait a minute—the sulphur—"

"Hush, Jules!" cautioned his uncle, as the lad showed signs of breaking in. "Let your brother finish."

"The sulphur," repeated Emile, his cheeks flushed with animation, "will stay on the surface; or perhaps it will sink, but not so fast as the iron, which is much heavier. "Let 's try it."

"I was confident, my good Emile," said his uncle, approvingly, "that you would soon get even with Jules. Yes, your idea is excellent, and if you hesitate a little in putting it into words, that is only because you are in some doubt as to how the sulphur will behave. I will put the thing to the test for you."

Uncle Paul thereupon took a large glass and filled it with water, into which he dropped a handful of the mixture, stirring the liquid at the same time with a small wooden stick. Having thus started a brisk movement in the glass, he paused and awaited results. Very soon the iron filings, because of their weight, had settled to the bottom, while the flowers of sulphur continued to circle about in the liquid. This liquid was next poured off into another glass, and when it came to rest the sulphur held in suspension gradually settled. Thus the two substances were collected, each by itself, the iron in the first glass, the sulphur in the second.

"You see, my young friends," said Uncle Paul, "it is accomplished quite as expeditiously as with the magnet, and the process calls for nothing that any one would not have at hand. Let us learn, I repeat, to do without what we lack and still to attain the end in view. It would be easy, you understand, for us to separate the two substances in the whole mixture by treating it a handful at a time in the manner just shown to you; but that is quite unnecessary for my present purpose. Let us sum up briefly what we have just learned. Two or more substances of different kinds form a mixture when their union does not prevent their being separated by the simple process of sorting, effected in one way or another. The heap there before you is a mixture of sulphur and iron, and these can be separated either with the help of a magnet or with water or, given sufficient time and patience, a grain at a time by hand. So much for that. Now let us pass on to something else."

So saying, he put the mixture of iron filings and sulphur into a bowl, added a little water, and kneaded the mass with his fingers until it formed a thick paste. Then he took a bottle of clear glass, an old discarded bottle that had once contained some sort of syrup or medicine, and filled it with the paste. Finally, in order to heat the mass somewhat, the bottle thus filled was set in the sun, and as it was a summer day the result purposed by Uncle Paul was not long in being attained, thanks to the temperature.

"Now pay close attention," he admonished his pupils, "and you will see something curious."

The boys were all eyes, all attention, in their eagerness to lose nothing of this their first experiment in chemistry. What was going to happen in the bottle? They did not have very long to wait. A quarter of an hour had not passed before something remarkable took place: the contents of the bottle, at first greenish in color from the yellow of the sulphur and the gray of the iron, began gradually to turn black and present the appearance of soot, while at the same time jets of vapor accompanied by hissing sounds escaped from the mouth of the bottle and small quantities of the black substance were ejected as if by the force of an explosion.

"Jules," said his uncle, "take the bottle in your hand a moment and, no matter what happens, don't loose your hold."

Unsuspectingly the boy approached and grasped the bottle firmly in his hand.

"Oh, wow!" he cried, with a start of pain and surprise; "it 's hot, hot!" And all his self-control was needed to prevent his dropping the burning bottle. Replacing it on the ground more quickly than he had taken it up, he turned to his uncle, shaking his fingers like one who has inadvertently touched hot iron. "How it scorches, Uncle!" he continued. "You can't hold it more than a second, it 's so hot. If the bottle had been over a fire I should have expected to find it hot; but there is no fire here to heat it, and yet it gets hot like that, all by itself! Who would have thought it?"

Emile in his turn had to handle the wonderful bottle that of its own accord grew so hot as almost to burn any one touching it. First feeling of it cautiously with his finger-tips, then grasping it boldly in his hand, he set it down again not less quickly than Jules had done, while his looks showed the profound astonishment, the utter bewilderment, caused by this generation of heat from no apparent source.

"Water was poured on the mixture of iron filings and sulphur," said he to himself; "it was all wet with water, which is not exactly the right sort of fuel for a fire, and then the whole was set in the sunshine, which is n't what you could call hot, and pretty soon, for no reason that I can see, the mixture grew scorching hot. I can't understand it."

Ah, my little lad, Uncle Paul's chemical experiments will give you many another surprise before they are finished! He who enters on the study of chemistry finds himself transported to a new world, where marvel follows marvel in endless succession. But don't be too bewildered; keep your eyes open, remember what you see, and gradually light will dawn on these perplexing operations which now seem rather to partake of magic than of veritable science.

"We have now learned," resumed Uncle Paul, "at the cost of some little pain to you, that the contents of the bottle become heated, apparently of their own accord, and that this heat is not slight, but very considerable, even sufficient to give a burning sensation. All the rest that happened we must regard as merely resulting from this development of heat. The water with which I moistened the mixture was turned to steam, and hence produced the jets of white vapor that escaped from the bottle. From this vaporized water came also the hissing sounds, the little explosions, and the throwing out of solid matter. If I had had at my disposal a larger quantity of iron filings and sulphur,—if my mixture, instead of being limited to a handful or two, had amounted to a full decaliter or more,—I could have produced some far more remarkable results. But I will content myself with describing to you a curious experiment that used to afford no little entertainment to the onlookers.

"A generous allowance of mingled iron filings and sulphur was placed at the bottom of a large hole in the ground, water was sprinkled over the mass, and a mound of damp earth was then heaped upon it. Soon this little mound would begin to behave exactly like a volcano in eruption: the ground would tremble all about the base of the mound, the heaped-up mass would crack open here and there, and through the cracks would spurt jets of steam accompanied by hissing sounds, explosions, and even tongues of flame. This was called an artificial volcano; but I must not omit to add that real volcanoes are set in action by something quite different from what was going on in that buried mixture of iron filings and sulphur, though this is not the time or the place to explain the difference. However, there is nothing to prevent your employing some of your leisure moments in constructing a miniature volcano of your own with a small quantity of iron filings and an equal amount of powdered sulphur. Your mole-hill of moistened earth, small though it must be, will not lack interest for you: it will at least break open in cracks and send out hot steam."

Emile and Jules resolved to gather up all the iron filings they could at the locksmith's and to buy a few sous' worth of flowers of sulphur, with which they would, at the earliest opportunity, perform the experiment of the artificial volcano. Meanwhile, as they were discussing this project, the agitation inside the bottle was gradually subsiding and the temperature rapidly falling, until finally the bottle became cool enough to be handled without inconvenience. Uncle Paul took it up and emptied its contents on a sheet of paper. What came out was a very black powder resembling soot.

"Now use your eyes," said he, "and see if you can find any of the sulphur; try to discover even one little grain of it if no more."

The boys rummaged through the heap, stirring it with a pin and scrutinizing it very closely, but could not point to a single particle of sulphur after all their pains.

"Where can it be now?" queried the searchers. "What has become of all that sulphur? It must be there somehow, for we saw it put into the bottle, saw it plainly enough. It is somewhere in that black heap; nothing could be more certain. It has n't been lost during the experiment, for it did n't come out of the bottle; nothing much except a little steam came out. It must be here, and yet we can't find the tiniest grain of it."

"Perhaps," suggested Jules, "we can't see it even if it is there because it has turned black; but we 'll try it with fire, and that will settle the question."

And convinced that he now had the solution of the mystery, Jules ran into the kitchen and fetched some live coals, on which he dropped a pinch of the black powder. But what was his disappointment when, after waiting a while and then blowing on the coals to make them burn more brightly, and after trying another pinch of the powder and then still another, each time from a different part of the heap, no ignition took place, no blue sulphurous flame showed itself!

"Well, I declare," exclaimed the bewildered lad, "that beats me! With all that sulphur somewhere in the powder, it still won't burn."

"And the iron," said Emile, "I can't see that, either. There 's nothing there but a sort of black soot, nothing at all that shines like iron. Let 's try the magnet and see if it will separate any of the filings from the rest."

But the magnet produced as little effect as had the live coals; no more bristling beards, no more strings of iron filings clinging to the poles of the magnet, after these had been passed to and fro through the black powder. Nothing was attracted, nothing showed any tendency to adhere to the piece of magnetized iron.

"Well, that 's strange," declared Emile, still pushing the magnet into the inert heap, now here, now there. "There 's plenty of iron there, that 's certain, and yet not a particle of it will stick to the magnet. If I had n't seen the iron put there I should say there was n't any in the whole heap."

"And I," chimed in Jules, "should say there was n't a particle of sulphur there, if I had n't seen it mixed with the iron. Yet of the two substances that certainly went into the heap, it now seems to contain not an atom; not a speck of sulphur, not a speck of iron can be found in what was made out of sulphur and iron."

Uncle Paul let his two nephews have their say, convinced that ideas thus born of personal observation are worth far more than those adopted on the authority of another. To see is to know. But after the boys had become thoroughly persuaded of their powerlessness to find and separate either the sulphur or the iron, then at last he intervened.

"Well," said he, "would you now undertake to sort the two substances, particle by particle?"

"It 's no use," was the reply; "we can't find the least trace of either of them."

"How about using the magnet?"

"That 's no good, either; it won't attract anything."

"Well, then, try water."

"I have n't much hope it will help us," answered Jules, "for the whole heap seems to be all of a kind, nothing heavy and nothing light. Still, it may be worth trying."

A pinch of the black stuff was dropped into water and stirred into the liquid, but it all sank very soon to the bottom of the glass, without the slightest tendency to any separation.

"So, then," resumed Uncle Paul, "sorting is no longer possible by any of the methods that at first succeeded so well. And that is not all: the appearance and the properties of the mass before us have undergone such a change that, if you did not know beforehand what was there, you would never suspect the presence of the two ingredients."

"But who in the world would ever imagine this black stuff was made of sulphur and iron?" the boys exclaimed.

"The appearance of the mass is changed, as I say," their uncle admitted. "The sulphur had a beautiful yellow color, the iron a lustrous gray, whereas the substance resulting from their combination is neither yellow nor gray nor lustrous; it is, on the contrary, of a deep, dull black. And the properties are likewise altered: the sulphur was found to take fire readily and to burn with a blue flame accompanied by stifling fumes, but this black substance refuses to ignite when it is placed on glowing coals; and the iron filings were attracted by the magnet, which has no effect on the black powder here. Hence we must conclude that this powder is neither sulphur nor iron, but some third substance of a wholly different nature. Shall we call it a mixture of sulphur and iron? Certainly not, for it is no longer possible to divide the mass into those two ingredients by any process of sorting, the properties of sulphur and of iron having given place to others showing nothing in common with the first two. We have, then, to do with an association far more intimate than that known as 'mixture,'—with one that is known in chemistry as 'combination.' Mixture leaves to the mingling substances their distinctive qualities intact; combination causes them to disappear, and substitutes others in their place. After mixture it is always possible to separate the ingredients by some simple process of sorting applicable to the given case; after combination this is never possible. Hence we may say that two or more substances are combined when they can no longer be separated by the process of sorting, in the customary sense of that word; when, in short, their characteristic properties have disappeared and given place to others.

"Observe, also, my young friends, that these new properties resulting from combination can by no means be predicted from the nature of the combining substances. Who would ever imagine, with no previous study of these curious things, that sulphur, yellow and readily combustible, could enter into the formation of a black and incombustible powder? And who would think that iron, with its metallic luster and its quick response to the magnet, could be capable of entering into the composition of a substance having a dull black color and no tendency whatever to be attracted by the magnet? Such things are impossible of prediction without previous knowledge. Combination, as you will have occasion to note again and again, works a fundamental change in matter, turning white to black and black to white, sweet to bitter and bitter to sweet, harmless substances to deadly poison and deadly poison to something entirely harmless. Watch well the result when two or more substances combine.

"Still another point demands serious attention. In the process of combining, our mixture of iron filings and sulphur became much heated by spontaneous action; in fact, it grew so burning hot that it was impossible to hold the bottle in one's hands. Jules will long remember the surprise caused him by this unexpected heat. In this connection I must tell you that this rise in temperature is nothing exceptional, nothing peculiar to the combining of iron and sulphur. Every time two or more substances enter into combination there is heat generated, sometimes so slight as to be detected only by the most delicate instruments; sometimes, and more often, of a degree unbearable to the touch; and sometimes, again, of such intensity as to be apparent to the eye in glowing redness or even blinding incandescence. In short, whenever combination takes place there is more or less heat; and, conversely, whenever heat or light is manifested it is almost always a sign that combination is going on."

"I should like to ask a question, Uncle Paul," Jules interposed. "When coal burns in a furnace, is there a combination going on between different substances?"

"Certainly there is."

"One of the substances, then, must be the coal, must n't it?"

"Yes, one is the coal."

"And the other?"

"The other is contained in the air. It is invisible, but none the less it is there. We shall consider it at length in its proper place."

"And the wood that burns in the fireplace and gives out heat and light?"

"There too we have a combination that includes the substance of the wood and that other substance contained in the air."

"And lamps and candles that we use for light?"

"Combination there also."

"Then every time I set fire to anything I start a combination?"

"Precisely; you cause two different substances to combine."

"What a funny thing it is, combination!"

"More than funny, my boy; it is useful beyond

your power to imagine, and that is why I wish you not to remain ignorant of the marvelous transformations it brings about."

"And will you tell us all about these wonderful things?"

"So far as I am able I will tell you about them, if you will both pay close attention."

"Oh, there 's no danger of our not doing that. We won't lose a word, and we 'll remember it all, too. I like this kind of lessons ever so much better than long division and conjugation of verbs. Don't you, Emile?"

"I should say so!" was the emphatic reply. "I wish I could have lessons like this all day and every day. I 'd leave my grammar any time to help make an artificial volcano."

"My dear young friends," their uncle admonished them, "don't let your enthusiasm for chemistry cause you to slight your grammar, if you wish to keep on good terms with me. Chemistry has its place, but so has language, and no small place, either. Don't neglect your conjugations, hard though they may seem to you. But now let us return to our subject of combination.

"It is, as I have said, always accompanied by heat, sometimes by light. Explosions, detonations, flashes of light, luminous outbursts, and brilliant sparks—all the dazzling display of an exhibition of fireworks, in short—are by no means exceptional when two substances come together in chemical combination. In the act of thus coming together the two substances unite in the closest of bonds; they marry, as we might say, and heat and light make haste to celebrate the nuptials just as pinwheels and Roman candles celebrate weddings with us. Do not laugh at my comparison; it is apter than you think. Chemical combination is like marriage: it makes one out of two.

"Now I have to tell you what this substance is that has resulted from the marriage of sulphur and iron. We cannot call it sulphur, as it is no longer sulphur; nor can we call it iron, as it is no longer iron. Neither would it do to call it a mixture of sulphur and iron, for what was a mixture in the beginning has ceased to be one now. Its name in chemistry is sulphid of iron, a name that enables us to remember the two substances united in the bonds of chemical matrimony,—iron, which we here write out unchanged, and sulphur, which appears somewhat disguised in the word 'sulphid.'"

## CHAPTER III

### THE SLICE OF TOAST

THE boys had made their little artificial volcano, and it had proved to be a success, the mole-hill of moist earth becoming much heated, cracking open, and giving vent to spurts of steam accompanied by sharp hissings. All had turned out to the complete satisfaction of the young experimenters, and the resulting sulphid of iron, on being examined at leisure and subjected to every test their imagination could suggest, was declared to be the same substance as that produced by their uncle. At this point he joined them.

"In the black powder now remaining at the heart of your artificial volcano," said he, "there is iron and there is sulphur. No possible doubt as to that can lurk in your minds after you have seen this substance prepared and, what is more, have prepared it yourselves with your own hands. Nevertheless, there is no sign of either any iron or any sulphur in this black powder, so utterly different is it in color and general appearance from both those substances. Had I begun by showing you this powder already made, without telling you of what it was composed, you would most certainly never have suspected it to contain any sulphur or any iron; and had I told you its ingredients without letting you witness the combining process, you would, I am sure, have taken your uncle's word for it, but at the same time you would have been no little astonished. 'What,' you would have exclaimed, 'sulphur in that stuff, there, which is not in the least yellow and will not burn? And iron, too, where there is n't the faintest shine of iron and nothing sticks to the magnet?' In short, you would have believed me because of your trust in my word, but you would not have had the certainty that comes from seeing the thing done.

"This certainty I have given you by means of the experiment performed before your very eyes, and you have further strengthened that certainty by performing the experiment yourselves. We are, then, all three of us, firmly convinced that in this black substance before us there are both sulphur and iron. And now another question arises: Is it possible to make the iron and the sulphur here combined resume each its original form? Can the combination be undone and the two ingredients recovered as they were in the beginning? Yes, my young friends, the thing is possible, but no simple process of sorting will suffice to disunite the two substances. You remember how all your attempts to accomplish this were so much wasted effort. What combination has joined together, no sorting can put asunder. To effect the separation, it is necessary to resort to scientific methods belonging to the domain of chemistry; and as your acquirements in that domain are still of the slightest, I will not invoke the aid of those methods. Besides, for our present purposes the actual separation of the sulphur and the iron is of very little importance. Inasmuch as the black powder does really contain them, it is incontestable that they can, by the requisite means, be obtained from that powder; and that is all I wish to impress upon you at present."

"There can't be any doubt," Jules assented, "that a substance made of iron and sulphur must furnish iron and sulphur when properly treated. No one could dispute that. All the same, I should like to see the iron filings come back as iron and the flowers of sulphur come back as sulphur."

"I repeat, my dear child, that the operation would not be difficult, but it would call for drugs quite unknown to you and would be a mysterious and perplexing performance in your eyes. Let us see but little at a time and see that little plainly; that is the way to acquire substantial and lasting knowledge.

"But, now that we are on the subject, I will say to you that what is done by combination is not always the easiest thing in the world to be undone. These chemical marriages, signalized by manifestations of heat and light, unite substances in bonds so close that to sever them it is necessary to employ methods known only to advanced science. However easy the act of union, the disunion is difficult. Combination takes place of itself; separation is a more arduous undertaking. We have lately seen the iron and the sulphur combine in a short time with no aid from us; but if now we should try to separate them, we should meet with enormous resistance, which only the most skilful methods could overcome.

"There are instances, however, in which quite the opposite is to be noted, combination being so difficult and delicate a process as to defy our utmost endeavors, but separation offering so little of resistance that a mere nothing, almost, will accomplish it. There are substances that dissolve their partnership with peculiar ease: a shock, a jar, a breath, an imperceptible trifle, will suffice to effect the severance. You touch them, you merely move them a little, and, *piff!* there is an explosion before you can snatch your hand away, with a flying of particles in this direction and that as if no such thing as union had ever existed. These are chemical marriages between incompatible natures that sigh only for divorce."

"And are there really," asked Emile, "substances that fly apart, that go *piff!* just from being touched?"

"Yes, my child, there certainly are. You yourself are familiar with some of them. Those New-Year's bonbons done up in particolored paper and known to you as snappers—don't they recall anything to your mind?"

"Why, yes; each bonbon has a rebus to be guessed, and then there 's a little strip of parchment that gives a pop when you pull it by both ends at once. What is it that makes the little explosion?"

"It is a substance made by combining different

ingredients which fly asunder as soon as they are disturbed by the parting of the two pieces of parchment forming the strip. You see how easy the act of separation is in this case: just disturb the slumbers of the explosive material by pulling at the two ends of the strip, and that is enough to cause a disruption accompanied by a sharp report. In like manner a house of cards collapses at a mere touch.

"A similar substance causes the explosion of the toy torpedoes that give a pop when you throw them on the ground, and to this substance is due also the explosive quality in the percussion-caps of guns, the cap being ignited by the fall of the hammer when the trigger is pulled. A quick spurt of flame is produced, and this penetrates the touch-hole and discharges the powder in the gun-barrel. Consider for a moment the construction of these percussion-caps. At the bottom of the little cup-shaped bit of copper forming the cap you can see a white substance deposited in a thin coating on the metal. It is the fulminating-powder, made of several ingredients carefully combined in accordance with chemical science and ready to fly apart with violence at the mere shock imparted by the hammer. But this is enough about these touchy and noisily dangerous substances, so prone to separate into their elements with a loud report as soon as we have joined those elements together. Let us proceed to something harmless. What should you say there is in a slice of bread?"

"I should say—I should say," Emile hastened to

answer, "that there is flour." And with that he thought he had said the last word on the subject.

"True," assented his uncle, "but what is there in flour?"

"In flour? What can there be in it except flour?"

"But what if I told you there was carbon, or what amounts to the same thing, charcoal, in flour?"

"What, charcoal in flour?"

"Yes, my boy, charcoal,—a lot of it."

"Oh, Uncle, you are only in fun! We don't eat charcoal."

"Ah, my young sir, you don't believe it? But did n't I tell you that chemical combination can turn black to white, sour to sweet, the uneatable to nourishing food? Furthermore, I will show you some of this charcoal that is found in bread; or, rather, I don't need to do that, as you have seen it hundreds of times and it will be enough now to jog your memory. Tell me: don't you often toast your bread a little over the fire before crumbling it into milk for your breakfast?"

"Why, yes, I let it get crisp and brown. It 's ever so much better that way; it goes better with milk when it is toasted just enough to make a crunching sound when you break it. In the winter, when the stove is hot, you can do it just right."

"But what if you forget your slice of bread on the stove? What if you let it toast too long? What happens to it then? Come, now, tell me, from your own remembrance of the thing, for I would n't on any account influence your opinion in this serious matter. What would happen if your bread stayed on the hot stove a whole hour?"

"That 's easy enough to answer: it would all turn to charcoal. I 've seen it happen lots of times."

"Well, then, tell me, where did the charcoal come from?—out of the stove?"

"Oh, no, not at all!"

"Then from the bread itself?"

"Yes, it must have come from the bread."

"But from no substance can there come anything that was not there before; nothing can furnish what it does not already have. Consequently, bread, which yields charcoal after being exposed some time to the action of fire, must itself contain charcoal, or carbon if we choose to use that word."

"Why, that 's so! I had n't thought of it before."

"There are many other things, my little lad, that you have seen again and again without grasping their significance, because no one has set you on the right road. I shall often turn these common occurrences to account by showing you to what important truths they open the way when you reflect on them a little. Reflection now makes you aware that bread contains quantities of carbon."

"I admit that bread contains carbon," assented Jules. "The proof is there before your eyes, plain enough. But, as Emile says, we don't eat charcoal, and we do eat bread; charcoal is black, and bread is white."

"If the charcoal, or the carbon, were alone," replied his uncle, "it would be black and uneatable, as you have described it, and it would remain so indefinitely. But it is not alone and by itself in bread; it is associated or combined with other things, and the combination has none of the qualities you have named as belonging to charcoal, just as sulphid of iron has none of the qualities belonging to sulphur and iron. These other qualities found in bread are driven out by excessive heat, and the charcoal remains, with all the characteristics peculiar to it,—blackness, hardness, brittleness, unpalatability,—in short, unmistakable charcoaliness. The heat of the stove undoes the work of combination, sundering what was joined together in the bread. That is the whole secret of the transformation of a slice of bread into a slice of charcoal when the toasting process has gone too far. Now let us inquire into the other things that accompany the carbon in white bread. They are known to you; you have seen them, and you have smelt their disagreeable odor when heat drives them out."

"I don't quite understand you," said Jules, "unless you mean that bad-smelling smoke that comes from bread when it is turning to charcoal."

"Exactly; you have my meaning. That smoke was a part of the bread it came from. The charcoal and the offensive fumes you know so well would, if recombined as they were at first, constitute precisely the slice of bread as it was before being subjected to the action of heat. Heat wrought the separation, dissipating some of the constituent elements in the air and leaving behind, stripped of its previous disguise, the black and uneatable substance so well known to you as charcoal."

"Then those bad-smelling fumes and the charcoal, with nothing else, make bread, and two things that could n't be eaten separately form by their union our chief food?"

"You have put it quite correctly: substances that by themselves, far from yielding nourishment, would be positively harmful if eaten, become by combination transformed into excellent food."

"I must believe you, Uncle Paul, because you say it is so; but—but—"

"I understand, my young friend, your hesitation and your 'buts.' On first hearing these things one can hardly believe them, so at variance are they with accepted notions. Therefore I do not ask you to take my bare word; you must be convinced by something other than my authority. Did I not at the very beginning prepare the way for these startling developments by means of a perfectly conclusive experiment? Recall the black substance that we obtained in the medicine-bottle. Recall that sulphur now no longer sulphur and that iron now no longer iron. Why should there be anything more surprising in the fact that charcoal and some bad-smelling fumes can cease to be what they now are, and can become bread?"

"You are right, Uncle, and the best thing to do is to take your word for it."

"To take my word for it sometimes may be necessary, as when the proof of an assertion would entail explanations too difficult for you to follow; but as far as possible I shall impose nothing on you as an article of faith, choosing rather to let you see, touch, and conclude for yourselves. I wish you to see the light and to witness the evidence, not to retain a mere mass of truths accepted on the authority of my word. In bread decomposed by heat I show you charcoal and call your attention to certain peculiar odors or fumes. What, now, is the natural inference?"

"That bread consists of that charcoal and those fumes united. It is too plain to be doubted."

"Yes, when facts speak we must accept what they say without heeding the counter-suggestions of long habit. These facts tell us that bread may be resolved by the action of heat into charcoal and certain vapors. Let us grasp that truth and acknowledge ourselves convinced.

"One other thing puzzles me," said Jules, "and it is the hardest puzzle yet. You say that the charcoal and the vapors separated by heat would, if recombined, make the bread again as it was before. Then, does n't fire destroy any of the bread?"

"The word 'destroy' has more than one meaning, my boy. If in using it you mean that a slice of bread, after being subjected to intense heat, no longer exists as bread, you are quite right: the resultant charcoal and vapors are in no sense bread, but merely the substances of which bread is formed. If, on the other hand, you mean that the bread is reduced to nothing, you are greatly mistaken, for there is not a particle of matter in existence that can by any force or device at our command be put out of existence."

"But that was just what I meant,—reduced to nothing, put out of existence. We speak of fire as destroying or annihilating everything."

"Then, in the literal sense of those words, we talk foolishly, for again I assure you that nothing in the whole universe, not even the tiniest grain of sand, is ever annihilated. Neither fire nor any other agency can annihilate even the finest thread of a spider's web.

"Listen, now, with close attention, for the subject is worth it. We will suppose a fine house is built, with spacious halls, splendid apartments, chambers, kitchen, vestibule, piazza, doors, windows,—in short, everything belonging to a comfortable and attractive abode. In building it the workmen had to place in their proper positions countless materials, such as cut stone, brick, rubble-stone, mortar, tiles, beams, boards, laths, plaster, metal fixtures, and so on. The house stands there, stanch and proud and suited to the requirements of the most exacting. Can it be destroyed? All too easily. Call back the masons with their picks and crowbars and hammers, and if necessary they will tear down the building much more quickly than they put it up. The fine mansion will soon be nothing but a shapeless pile of ruins, of rubbish; it will be destroyed as a house.

"But will it be annihilated, reduced to nothing? Evidently not. Does there not remain an enormous heap of materials,—of stone, brick, wood, iron,—

of everything, in fact, that went to the building of the house? The house, then, is not annihilated, and, what is more, not a particle that entered into its construction has been reduced to nothing. Even the last grain of sand used in mixing the mortar is sure to be in existence somewhere. The wind may have blown away some of the plaster-dust as the house was being torn down; but that dust, of a fineness hardly visible, is nevertheless undestroyed, however widely dispersed by the wind; and if it cannot now be gathered up, we can at least see it in our mind's eye, scattered in this direction and that. Of the entire building, therefore, that has been demolished not a particle of dust has been annihilated.

"Well, now, fire in its turn is a demolisher, but nothing more. It demolishes buildings made of many materials combined, but it never reduces to nothing the smallest particle, the minutest grain of dust, in those materials. We subject to its destructive power a mouthful of bread, and destruction follows, but never anything like annihilation; for what is left, after the fire has played its part, is just as truly matter as was the bread itself. That residue is in the form of charcoal and certain fumes or vapors, the charcoal remaining in a little mass by itself, the vapors being dissipated and no longer traceable, even as the plaster-dust was lost to view. Rid yourselves, then, forever, of the foolish notion of annihilation."

"But—"

"There goes Jules again with another of his

'buts'! What is your difficulty this time, my lad?"

"When you burn a stick of wood in the fireplace, isn't it reduced to nothing, or almost nothing? There 's only a pinch of ashes left at the end. I see how the ashes come from what was once wood, but they amount to so little they can't represent all that has been demolished by the fire. The greater part of the wood, then, must have been reduced to nothing."

"Your observation shows a thoughtful mind, and is of the kind I like. Accordingly, I hasten to answer you. I just spoke of the plaster-dust blown away by the wind in the demolition of our supposed house. Is it not plain that, the walls being built largely of powdery materials capable of being caught up by a passing breeze, a considerable part would be thus borne away in various directions, leaving behind a proportionately diminished heap of refuse?"

"Certainly; I admit that."

"If, now, it were possible in a work of masonry for the whole structure to be swept away as impalpable dust, what would remain?"

"Nothing, of course."

"But would the building on that account have been reduced to nothing?"

"Why, no; it would have been turned into fine dust scattered all about."

"Just so with your stick of wood, my little friend: fire resolves it into its constituent elements, some of which are far more impalpable than the finest dust. These are lost to view, being dissipated here and there in the boundless atmosphere, and as we find nothing left but a handful of ashes we are prone to believe the rest has been annihilated, whereas it still exists, indestructible, floating in the atmosphere and having a limpidity, a colorlessness, as complete as that of the air itself."

"Then a stick of wood that has just been burnt up in the fireplace is mostly scattered in the air in a sort of fine dust that we can't see?"

"Yes, my boy; and the same is to be said of all fuel that we burn to obtain either heat or light."

"Now I see why wood, when it is burned, seems to be reduced to nothing. What was the wood has, as you say, been mostly carried away without our seeing it, somewhat as the plaster-dust of a house that is being torn down is blown away by the wind."

"Note also, my boys, that out of the materials left when a house is torn down, another house can be built, different in form and on another site if desired. The heap of ruins will thus become once more a finished structure. But, further, there is no reason why these same materials could not be used for making other things, the stones for one purpose, the bricks for another, the wood for still another, so that the ruins of our demolished house would enter into various constructions having each its own form and purpose and character.

"Somewhat thus is it with matter in general. Let us suppose two, three, or four substances, each of a different nature from the others, to enter into combination. They function all together in a certain manner; they dispose themselves so as to form what

I will call a kind of building; and by thus associating they produce a substance quite different from any of the constituent substances, just as our finished house is neither sand nor lime, nor plaster, nor brick, nor, in fine, any of the materials used by the builders.

"After a while, for some reason or other, these combined substances separate, and the chemical structure is demolished. The ruins are left; there has been no loss of matter. What will nature do with these ruins? Perhaps any one of a thousand things; perhaps use a little of this ingredient for one purpose, a little of that for another, and so on until the last particle of matter has been utilized and the result is a great variety of productions, all very different from the original substance. What went to make something black, will, it may be, now enter into the formation of a white substance; what was a part of something sour, may contribute to the making of something sweet; and what helped to constitute a poison, is likely enough to be found again in an article of food, just as the bricks of a former conduit may by a totally different application serve in the construction of a chimney and thus make a passage for smoke and flames instead of for water.

"Thus it is that nothing is ever annihilated, despite all appearances to the contrary, appearances that so often deceive us because we do not observe accurately. Let us pay closer attention, and we shall perceive that all matter persists, indestructible. It enters into an infinite variety of combinations, forever uniting and separating and uniting again, some of its manifold forms being every moment destroyed and every moment renewed, in an endless series of transformations, without the loss or gain of a single particle in the whole universe."

## CHAPTER IV

### SIMPLE SUBSTANCES

"LET us return now to the black powder, the sulphid of iron, that served as our starting-point in this discussion. By a process far less simple than ordinary sorting, a process known to chemical science, this substance can be decomposed and the sulphur and iron separated. Subjected to the easy decomposition wrought by fire, bread furnishes, as its most notable constituent, carbon. Now of what, in their turn, are carbon and sulphur and iron each composed? Let me give you the answer made to this question by scientific investigation as conducted ever since these substances became the objects of man's interest and study. No matter with what thoroughness, with what elaborate and painstaking experiments, they are examined, no matter how powerful the forces brought to bear on them, carbon and sulphur and iron never give us anything but carbon and sulphur and iron."

"But it seems to me," objected Jules, "that sulphur does give something that isn't sulphur. When you set fire to a little of it, there is a blue flame and some sort of vapor that makes you cough. That vapor must come from the sulphur, but it is something very different from sulphur, for it makes us cough worse than if we had the whooping-cough, and sulphur does n't do that even if you hold it right under your nose."

"Let us understand each other, my boy. When I say that sulphur never gives us anything but sulphur, I mean that it cannot be decomposed into other substances; but I do not by any means assert that it cannot by combination with other substances produce not only the vapor that makes us cough, but also many other things, notably the black powder you now know so well, the sulphid of iron. I told you that every substance, in burning, combines with another which we cannot see, and which is contained in the atmosphere about us. If sulphur becomes enveloped in blue flames, it is a sign that it is combining with that atmospheric substance. The result of this combination is the vapor that makes us cough."

"Then that vapor is more complicated than sulphur?"

"Yes."

"It must be made of two things, sulphur and that stuff in the air you told us about, while sulphur is made of only one thing, sulphur itself."

"Quite right. I repeat, then, that sulphur, even when put to every sort of test, has never been decomposed, never been divided into different substances, as, for example, the black powder in our medicine-bottle could be divided into iron and sulphur, and as bread could be resolved into several ingredients, carbon among them. Sulphur goes to the making of a great many things more complicated than itself, but can never yield anything simpler. When we come to sulphur, decomposition stops; by no means at our command have we yet succeeded in dividing sulphur into two or more other substances. Hence we call sulphur a simple substance, meaning that its further simplification is impossible. Water, air, a pebble, a piece of wood, a plant, an animal—all these may be regarded as substances, but they are not simple substances. Bear that in mind.

"Carbon and iron are also simple substances, for the same reason that sulphur is a simple substance: they cannot be made to yield anything except carbon and iron unless they are combined with other things; but this would not be a simplifying, it would be a complicating process. Chemists have carefully examined all substances to be found in nature, whether on the earth's surface or beneath, whether in the depths of the sea or in the air about and above us, and whether belonging to the animal or the vegetable or the mineral kingdom; they have examined all, studied all, analyzed all, and the fruit of this immense task, prosecuted with all possible learning and patience, is the conclusion that the number of undecomposable or simple substances amounts to sixty[1] or thereabout, including iron, sulphur, and carbon, which we have just been considering."

"And will you tell us about all these simple substances?" asked Emile.

[1] Since this was written twenty or more have been added to the list.—*Translator.*

"Not all; far from it, for the greater part would not interest us; but you shall hear about some of the more important ones. Moreover, you already know a number of simple substances besides the three—iron, sulphur, and carbon—which we have just examined."

"I know other simple substances?" the boy exclaimed in surprise. "I did not know I was so wise!"

"You knew carbon without suspecting it of obstinately resisting all attempts to decompose it. There are more things in your good little head than you are aware of; my part is to put some sort of order into your jumbled ideas. But I shall refrain as far as possible from teaching you outright, preferring to make you recall what you already know. I will tell you at this point, however, that all metals are simple substances."

"I see. Then copper, lead, tin, silver, gold, and others that I forget, are simple substances, just the same as iron."

"An expert chemist could not have put it better. They are so many substances on which decomposition has no hold; they are simple substances. But there is one metal in common use that Emile has omitted. Think! It begins with *z*."

"With *z*? Wait—it is zinc, the same as our watering-cans are made of."

"Right. These are not nearly all the metals, though; there are many others, and among them some extremely curious ones, but they are not in general use. I will acquaint you with them as fast as we have occasion to handle them. One, however, might be mentioned here. This metal runs like melted tin, but at the same time it is cold. It has the color of silver, and it goes up and down in a thread-like column in the thermometer to show us the temperature of the atmosphere."

"Oh, that is mercury, or quicksilver!"

"Exactly. Its common name of quicksilver might deceive you. It has the shiny appearance of silver, but none of that metal's other qualities. It is a distinct metal, as different from silver as are lead and copper. The term 'quick' indicates that it runs all about, flowing in little globules and always eluding the fingers that try to grasp it."

"Then quicksilver is a metal, the same as iron, copper, lead, or gold?"

"It is a metal, no more and no less, but different from others in that the mere warmth of our climate, even in winter, is sufficient to keep it in a molten state, whereas to melt lead the heat of burning coal is necessary, and for copper and especially iron it takes the hottest kind of furnace. But were it cooled sufficiently, it would become hard and not unlike a piece of silver in appearance."

"Money could be made of it, then?"

"There would be nothing to prevent; but it would be a strange kind of money, for a few minutes after putting it in your pocket you would find it melted and running all about.

"The color of metals does not vary much: silver and mercury are white, tin a little less so, and lead still less, while gold is yellow, copper red, and the others, notably iron and zinc, grayish white. All shine brightly, at least when recently cleaned; or, in other words, they all have a metallic luster. But you know the proverb, 'All that glisters is not gold.' In the same way, all that shines is not metal. You would not have to hunt long in the garden to find some insect—some beetle, for instance—whose rich wing-sheaths have the luster of polished metal, though in reality they are nothing but scales of horn. Certain stones, judged by their deceptive luster, might be mistaken, some for gold, others for silver; and yet they contain not the slightest particle of these two metals. The glittering yellow spangles [1] mixed with the blue sand used for drying ink after writing have nothing in common with gold except the glitter, and are not even made of metal. So it is that all metals, without exception, have the peculiar luster called metallic; but this luster may be found, just as bright, in a great many other things that are not metals.

"The other simple substances, sulphur and carbon among them, are without metallic luster; some there are, too, and very important ones, that are colorless, invisible, of the same subtle quality as air. These non-metallic simple substances are called by the general name of metalloids. Carbon is a metalloid, sulphur is another. The number of metalloids is not great, about a dozen perhaps, but the part they play is highly important. One might say of them what is, alas, only too true of human

[1] Particles of mica. Before blotting-paper came into use, fine sand was sprinkled over fresh writing to absorb the ink.—*Translator.*

beings: those that make the greatest noise are not the ones that are the most useful. Indeed, though the metalloids are of primary importance in the construction of the things about us, being no less necessary in the countless works of nature than stone and bricks and mortar in our buildings, there are some that many of us do not even know by name. One must at least have read some book on chemistry to be aware of their existence. If we were not told of them by those more learned than ourselves, we should remain ignorant of them to the end of our days. One of these important substances—and it is one without which we should speedily die—has a name that is probably unfamiliar to your young ears; you may, in fact, never have heard it mentioned. It is oxygen."

"Oh, what a funny name!" cried Emile. "I 've never heard it before."

"And these: hydrogen, nitrogen—do you know them?"

"No more than the other."

"I suspected as much. They are the names of two good and useful metalloids that quietly perform their appointed tasks without soliciting public notice, just as a generous giver is willing to remain unknown, provided only his gift reaches its destination. All three—oxygen, hydrogen, and nitrogen—are the less likely to attract popular attention, despite the importance of their services, because they have the invisibility, the thinness, of air. Very often, too, they are concealed in combinations in which only the higher science can detect their presence. Reason enough, then, for our remaining in ignorance about these substances that play the leading parts in nature's never-ending drama."

"Are they, then, so very important?"

"Yes, my boy, they are extremely important."

"More so than gold?"

"You are all astray, my dear Emile, on this subject of importance. Gold is unquestionably a very useful metal to man; it is the sign of riches, of the savings amassed by labor. Coined into money, it passes from hand to hand and is good in all exchanges, in all commercial transactions. It is a splendid part to play, I admit; but if gold were to disappear entirely from the earth, what would happen? Nothing very serious. Banks might be inconvenienced, commerce upset for a little while, but that is all. The world would soon move on again as before. Suppose, on the other hand, one of these three metalloids whose names you have just learned—oxygen, for example—should disappear. Immediately everything on earth would die, from the biggest animal to the tiniest worm; all plant life would perish, from the giant of the forest to the smallest thread of moss. Life would henceforth be impossible and this inhabited globe become a gloomy solitude, with man, animal, and plant forever banished. That, as you see, would be a far more serious disaster than the inconvenience of a banker or the vexation of a merchant.

"In the general scheme of things, gold plays only an insignificant part, almost a negligible part. If it were lacking altogether, the order of nature would not be affected. Oxygen, hydrogen, and nitrogen, on the contrary, fulfil in this world of ours functions so important that if any one of these three were taken away, everything would be turned topsy-turvy and life would be rendered impossible. To these three carbon must be added, for its part is not less important; and thus we have four substances indispensable to all life, vegetable as well as animal. Now compare with them, if you like, this gold that everybody talks about, is familiar with, longs for, and that many wear themselves out in trying to get. Was I not right when I said that making the most noise in the world is a very different thing from rendering the highest service? Believe me, my young friends, gold is but a poor thing when looked at from the proper point of view."

"And you will tell us about this oxygen, this hydrogen and nitrogen, that the world cannot do without?" asked Jules.

"Certainly. I shall begin with them. Honor to whom honor is due. To tell you what they are and what they do will take up more time than all the rest. To complete the list of metalloids that you need henceforth to know, at least by name, I will mention one other. It is the substance you see at the tip end of a match, overlying a layer of sulphur, and taking fire with friction. It will give a mild glow, too, when rubbed between the fingers in a darkened room."

"That must be phosphorus."

"Yes, phosphorus. It also is a metalloid. Let

us sum up at this point. There are about sixty simple substances, divided into metals and metalloids. Metals have a peculiar sheen called metallic luster. Those known to you are iron, copper, lead, tin, zinc, mercury, silver, and gold. Some of the others deserve our attention as well, and I will speak of them as occasion arises. There are, all told, about fifty[1] metals. The metalloids, which are much fewer in number, or about a dozen in all, have not the so-called metallic luster. The most important ones are oxygen, hydrogen, nitrogen, carbon, sulphur, and phosphorus. The first three in the list are, like air, invisible.

"Simple substances, metals as well as metalloids, are also called elements. By this is meant that they are the undecomposable or prime substances used by nature in all her works."

"But, Uncle Paul," Jules here interposed, "I read in a book that there are only four elements in nature, not sixty, and that they are earth, air, fire and water."

"That book repeated the false notions of ancient times, notions preserved to our day in popular speech, in which long habit is slow to give place to the progress of science. It was, indeed, formerly believed that everything in nature could be traced back to earth, air, fire, and water, which were thought to be the four undecomposable substances, the four elements, of which all things were made.

[1] The recognized number is now nearer seventy than fifty.—*Translator.*

But it has been found after more careful study that not one of the elements as understood by the ancients is really a simple substance.

"In the first place, fire—or, rather, heat—is not a material thing at all, and consequently has no place in the list of simple substances, which are matter even when they are invisible. All matter can be weighed and measured. We speak of a cubic foot of oxygen, a pound of sulphur; but it would be the height of absurdity to speak of a cubic foot of heat or a pound of warmth. You might just as well pretend to weigh by the pound and measure by the quart the notes that come from the strings of a violin."

"A pound of F sharp or a quart of E flat would certainly sound funny," Jules agreed, smiling at the odd association of words.

"You smile, naturally enough, for a musical note cannot be weighed in the scales or measured by the quart or the bushel. And why is this impossible? Because sound is not matter, but a movement transmitted in successive waves from the sonorous body to our ears. The same with heat: it is a peculiar mode of motion. To my regret, I can but touch on this interesting subject at this time, since to explain it properly would take so long that chemistry would in the meantime be forgotten. I will simply say, then, that heat cannot be classed as an element, because it is not matter.

"When we come to air, however, we have quite another thing. It can be measured by the quart, weighed by the pound. It is probably new to you to hear air spoken of as being weighed and measured; but nothing could be more correct. Physics could teach us much about this if we had the time just now to give to it. Air is matter, though, unfortunately for the credit of the ancient theory, this matter is not a simple substance. Instead of being made of one thing, air is made of two very different things. I will tell you their names before I undertake to prove to you by experiment the truth of this assertion. Air is composed of oxygen and nitrogen.

"Nor is water any more truly a simple substance, an element, than air. At the proper time I will show you that it is a compound of oxygen and hydrogen.

"As to earth, what is meant by that word? Evidently the mixture of mineral substances—sand, clay, gravel, pebbles, rocks, stones—that form the solid part of the globe. Thus, instead of being an element, one simple substance, it contains all the elements, from the first to the last. From the earth are obtained all the metals and various metalloids; in fact, all simple substances could be derived from this source if we cared to decompose those combinations in which many of them are at present tied up. We may, then, say of the four elements, commonly so called after the ancient conception of the matter, that not one of them will bear critical examination, not one will prove to be a simple substance, an element, as the word is understood today."

## CHAPTER V

### COMPOUND SUBSTANCES

"WITH the materials of his handicraft—with stones, bricks, mortar, and plaster—a mason can build, at his will, vault, bridge, wall, reservoir, shed, coach-house, factory, cellar, terrace, hut, castle, or palace; and each of these constructions, although of like materials, will differ in form, purpose, and other qualities. In similar manner, with the threescore elements at her disposal, nature fashions all the things that come from her hand, in the animal, the vegetable, and the mineral kingdoms. An artisan of sublime achievements, she demands but a few materials; and not even all of these does she often use at once, in order to obtain results of infinite variety. Combined in countless ways, these elements or simple substances form everything on and in the earth. Nothing, absolutely nothing is known to us that, when decomposed (if it be not simple to begin with), fails to resolve itself into a certain number of metals or metalloids or both."

"Then everything comes from these same simple substances?" asked the children.

"Everything that is not already a simple substance. Consider for a moment the element that you see most often under one or another of its various disguises,—carbon. I have shown it to you as forming a part of bread. You know, too, that wood contains it, as can be seen from the charred fagots in the open fire. Now, the carbon in bread and that in the trunks and branches of trees are exactly the same, so that in the ever-changing combinations of nature the carbon in the loaf of bread might reappear in an oak fagot, and that in the oak fagot might turn up again in the loaf of bread."

"And so," observed Emile, more in jest than in earnest, "when we eat a slice of bread and butter, we are eating what might have made a knotty stump."

"Who knows, then, my little lad," his uncle took him up, "how many knotty stumps you have eaten in your lifetime? I hope soon to show you that your jesting remark comes nearer the truth than you thought."

"Uncle Paul, I won't say anything more! Your simple substances are too much for me."

"Too much for you? Not at all. But perhaps you feel yourself, for the moment, a little dazzled by the blinding light of a new truth, just as a strong ray of sunshine dazzles the eyes. Let us continue, and gradually everything will appear clear to you. The carbon in an oak fagot—why should it not have gone to the making of a pear, an apple, or a chestnut? Is there not carbon in those?"

"Yes, there is," replied Jules. "When you leave chestnuts on the coals too long they turn to charcoal, and if you forget apples or pears put to bake in the oven you find them nothing but lumps of charcoal."

"Again, this charred chestnut, apple, or pear is of the same substance as that in firewood and in bread. Do you now begin to glimpse the fact that it is possible to eat what, by a change of destination, might have become a stump or a stick for the fire?"

"I can more than glimpse it," Emile answered; "I see it."

"And you will soon see it clearer. If instead of using olive-oil on our salad or for frying fish, we put it into a lamp, it will burn and give light. Now let us hold a piece of window-glass or a plate over the flame. Instantly a coating of black dust collects on it."

"I know, that is lampblack. I make my glasses dark with it when I want to look at the sun in an eclipse."

"And what is this lampblack?"

"It looks very much like charcoal-dust."

"It really is charcoal, or carbon. And where does this carbon come from, if you please?"

"I don't see where it could come from, unless from the oil burned in the lamp."

"It does come from the oil, that is plain; from the oil decomposed by the heat of the flame. So there is carbon in oil. Needless to add, this carbon differs not at all from other carbon. It is found in grease, in tallow, for candles and tapers give lampblack just as does the oil-burning lamp. It is also in resin, which burns with a thick black smoke; it is in— But I should never get to the end if I tried to give a complete list. I will mention finally the mutton-chops you have seen so often on the dinner-table. If the cook is not careful, what becomes of them on the gridiron?"

"Why, that 's so!" Emile exclaimed. "I had n't thought of it. If you let them cook too long, the chops all turn to charcoal."

"What, then, do we infer from that?" asked Uncle Paul.

"We infer that there is carbon in meat. It must be everywhere!"

"Everywhere? Oh, no! Far from it. But carbon occurs very often. You will find it especially in all animal and vegetable products. All these substances, when decomposed by fire, leave carbon in their ashes. So you can easily make out as long a list as you choose of substances containing this element."

"Paper, white as it is, must have some, for it turns black when you burn it. But, tell me, does paper come from plants?"

"Yes, my child, it comes from vegetable matter, being made out of old rags, and these old rags are the remnants of fabrics woven of linen or cotton."

"Milk," asked Jules, "which is still whiter than paper—does that too have carbon in it? I have seen the foam turn black at the edge of the saucepan when the fire was too hot."

"Yes, milk too contains carbon, I can assure you. But let that do for the present. Further examples are not needed to show you what varied uses carbon can serve at the hands of Mother Nature. Now will

Emile recite the fable he has been learning by heart the last few days?"

"Which one?"

"The one about the sculptor and the statue of Jupiter."

"Oh, yes, I know:

"A block of marble, fair to see,
Filled with delight a sculptor's soul.
He bought it. 'Now what shall it be,
A god, a table, or a bowl?

" 'A god were best, an awful god,
With thunderbolt in lifted hand.
Mankind shall tremble at his nod,
His name be feared in every land.' "

Here Uncle Paul stopped the reciter: "That is enough, my boy. You have a good memory. What does the good La Fontaine tell us? He tells us that a sculptor, viewing the superb block of marble he had just bought, asked himself what he should do with his purchase. His chisel could make of it at will a bath for a sumptuous palace, the bowl of a fountain for a princely garden, or a modest slab, a commonplace bureau-top or mantelpiece. He decided on a god. The block of marble which could become a bowl in which to wash face and hands, shall be Jove the Thunderer, before whom all mankind falls prostrate. Out of one and the same material the chisel is to bring forth, not a trivial piece of furniture, but a noble statue. In like manner does Nature proceed, able as she is to make whatever she chooses out of the chemicals at her disposal.

A little carbon, let us say, is at hand. 'What shall my art make of it?' she asks. 'Shall it be a flower, turnip, flesh, or hair of an animal? It shall be a flower; more than that, it shall by its coloring and perfume be the queen of the flower garden.' And the splendid rose comes forth from the carbon that might have become sheep's tallow or a part of a donkey's horny hoof."

"But there 's something else in the rose besides this carbon that makes so many things, is n't there?" asked Emile.

"Certainly; otherwise the carbon would remain carbon and nothing more. It is combined with other simple substances. The same must be said of it in all the other things we have just named as containing carbon."

"Then," said Jules, summing up what his uncle had told them, "bread, milk, grease, oil, fruit, flowers, linen, cotton, paper, and lots of other things, all contain carbon and also various other elements that never change their nature, whether they are in a flower or in a lump of tallow, in a piece of paper or in a stick of wood. They are always the same metals and the same metalloids. And are our bodies too made of these things?"

"As far as matter is concerned, man does not differ from the rest of creation. His body has for its constituents exactly the same metals and metalloids."

"What!" cried Emile, surprised at this human chemistry. "Are there metals in us? Are our bodies mines? I could believe it if we were all sword-swallowers like the jugglers we see at the fair; but we have n't quite come to that yet."

"Agreed. Nevertheless, there is iron in us,—precisely the same metal as is swallowed by the jugglers you speak of. And iron is so indispensable to us that without it we should lead a languishing life; or, rather, we should find it impossible to live at all. It is iron, let me add in this connection, that gives to our blood its red color."

"I know that our blood is colored red somehow or other; but, all the same, I know that nobody can eat metals, not even the juggler who fools us with his cleverness. Where does this coloring matter come from, then?"

"Like carbon, sulphur, and whatever other elements the body needs, it comes from our food, which contains it, a little here and a little there, without our knowing it. And are you quite sure that we never take iron, real iron without any disguise? At your age, when the mere work of growing is something considerable, and strength is none too great, the doctor often orders iron, which is taken in the form of a very fine powder, or is given to us to drink in water that has had old iron in it for some time, and has thus become slightly charged with the rust. That is not exactly the same as swallowing swords, but it is eating iron nevertheless."

"I am ready now to believe we eat as many metals as you please," Emile assented.

"Not so fast! Don't let us make the human body a mine, as you called it just now. I am speaking only of iron, to which might be added three or four other metals that, unfortunately, you know nothing about as yet. There are metals familiar to us all, such as lead, copper, zinc, gold, and silver, that have no place in the human body or in animals or in plants. Certain metals, indeed, if they were introduced into the body, would endanger life, for they are poisons. I keep to iron, then, and will add that very little of it is enough to give color and other peculiar properties to the blood,—so little, in fact, that the body of an animal the size of an ox would furnish hardly enough iron to make a nail. I will add that this nail would cost a fabulous sum, so much labor and pains would have to be expended in the mining of this animal ore. If necessary, the thing could be done, which is all I wish to make you understand.

"We have now reached the point where you ought to begin to perceive that simple substances, by combining in many ways, produce a vast variety of other substances endowed with widely varying properties. These substances are called compound because they are each composed of a number of elements. Water is a compound substance; so are flour, wood, and paper, oil and grease, pine resin, animal flesh and horn, the essence of the rose, and, in short, such a multitude of things that the list would never come to an end. Water is composed of oxygen and hydrogen, two metalloids that we shall take an early opportunity to become acquainted with. The other things I have just named contain carbon among their elements.

"So vast is the number of compound substances

that we might almost call it infinite. At any rate, no limit is known. And yet all these compounds come from the mixing together of two or more of those simple substances that do not number so very many, some threescore in all. Furthermore, many of these simple substances play so unimportant a part that their entire omission would make no appreciable difference in the grand total of the world's material riches. Among these minor elements I will mention gold. Confining ourselves to the fundamentals, we can see that, at most, only about a dozen simple substances contribute to form the immense majority of the products of nature."

"But there 's one difficulty, it seems to me," objected Jules; "and it strikes me all the more after what you have just said. I was wondering how such a lot of different things—so many that perhaps they could n't be counted—can all come from sixty elements. And now I wonder still more how the great majority of these countless things can come from only a dozen elements."

"I was expecting this objection to be raised, and was about to answer it in advance when you got ahead of me; for which, in truth, I am very glad, as thus I receive a fresh proof of the reflective quality of your mind. I have proposed a puzzle to you, and now I will use an illustration that will help you to solve it. Our alphabet has twenty-six letters. How many words can be formed with these characters?"

"Why—I—I don't know what to say. I have never counted them, for a dictionary, even a small one, has lots and lots of words. Let us say ten thousand."

"We will let it go at that—ten thousand, in round numbers. It is not necessary to be very exact in this matter. You will notice that we are speaking only of our own language; but the same characters could be used for writing all the languages of the world that have been spoken in the past, that are spoken to-day, or that may be spoken in the future. I omit certain instances of peculiar pronunciation, which are negligible in this connection. With our twenty-six letters, then, Latin, French, English, Italian, Spanish, German, Danish, Swedish, and many other languages are written. The same letters, too, could be used for Greek, Chinese, Hindustani, Arabic, and all other tongues with written characters differing from ours only in form. There is no language, even to the lowest negro dialect, that could not be represented in some sort by our alphabet. In this grand total of languages and dialects what multitudes of words there must be!"

"We should have to count them," said Jules, "not by tens of thousands, but by millions and millions."

"Now, imagine for the moment, my boy, that these letters represent our simple substances, while words represent the compound substances. The comparison is not so very far-fetched, for just as words having each its own value, its peculiar meaning, are formed by combining letters in groups of two, three, four, or more, and in such and such an order, so compound substances are made by combining certain elements which, according to their properties, their number, and the manner of their grouping, determine the nature of the compound."

"Simple substances, then," put in Jules, "are the elements of material things just as letters are the elements of words."

"Yes, my boy."

"Then the number of compound substances must be as immense as the number of words in all the languages of the world. Still, I should say the alphabet would give the greater variety. It has twenty-six letters, and you have just told us that most compound substances are made from a dozen elements at most. Twenty-six ought to give more combinations than twelve."

"I will ask you to note that the number of letters might be considerably reduced and still the alphabet would represent all the various vocal sounds. What difference is there, I ask you, in the pronunciation of *k, q,* and hard *c*? None. One of these characters is necessary, the others are superfluous. In like manner soft *c* is the same as the hissing *s,* and *x* is simply *ks,* nor does *y* as a vowel differ from *i.* Rid of its duplicating characters, the alphabet could, as you see, be reduced a good deal and still be rich enough to furnish the elements of innumerable words. But I admit that even so there remain more letters than there are simple substances forming the great majority of compounds. In their modes of grouping, however, the elements enjoy a great advantage over the letters of the alphabet.

"To make a word, we usually group two, three, four, often five or six, and even more letters. Take for example, that long and cumbrous word, *inter-communicability.* One must draw a good breath in order to pronounce it all. There are twenty letters in the word, or almost as many as there are in the whole alphabet, though it is true that some of these letters are repeated, which reduces the number of separate characters to thirteen. Chemical combination scorns such cumbrous piling up of elements, and imposes upon itself a rigid rule never to resort to it, holding that complicated mixtures are none of its business. To form compound substances it groups only two simple ones, or sometimes three, very rarely four. Imagine a language with words of only two, three, or at most four letters, and you will get a notion of the compound substances resulting from the union of chemical elements. Sulphid of iron is a compound of two elements,—a word of two letters, if you choose, continuing our comparison. Water is another. Oil has three, and animal flesh four. Compounds of two elements are called 'binary compounds'; those of three, 'ternary,' and those of four 'quaternary.' These terms come from the Latin words meaning respectively two, three, and four.

"Now, if four elements at the most, and commonly only two or three, are united in combination, how is it that there can be an almost infinite variety of compounds? To aid us in explaining this, take the word *rain,* for example. For the initial letter *r* substitute another, and then another, and so on, and we have the common words, *gain, lain, vain, wain, pain,* and others, all belonging to our language. In the same way *pin* becomes *tin* and *din* and *sin*. By a simple change of one letter, the rest of the word remaining unaltered, we have a word of a wholly different meaning. So it is with chemical compounds: let one element be replaced by another, the rest remaining unchanged, and behold, at once we have new properties, a substance very different from the first.

"But, further than this, there is still another change that gives an even greater variety of compounds. Just as in a single word the same letter may be repeated several times (note the letter *i* occurring four times in the long word just cited as an example), so the same element is, in many substances, repeated in chemical combination. It is taken two, three, four, five, and even more times, producing each time a compound having its own peculiar properties. We should hunt in vain in the dictionary for words suitable as illustrations of this principle, for our language refuses to repeat the same letter over and over again in one short word. But let us imagine a series of words such as *ba, bba, bbba, bbbba,* and so on, and let us suppose that each of these, although containing only the letters *b* and *a,* one of which is repeated, has a meaning entirely different from that of any of the others. We can thus gain a fairly good idea of what takes place in compound substances."

"If that 's the way of it," said Jules, "I can see well enough that the number of compounds must be very great,—that it must be enormous, even with only the dozen simple substances that play the chief part. One element changed and another repeated must produce an almost endless variety of different groups."

"And what does Emile think about it?" asked his uncle.

"I rather agree with Jules: there 's much more variety than I had thought. But I should see it better if I could understand how *bba* is really different from *ba.*"

"You would like to have an example of a compound substance whose nature changes completely when one of its elements is doubled?"

"Oh, Uncle, that 's just what I should like to see; and Jules would too, I 'm sure."

"I can easily gratify you, my little lads."

And so saying Uncle Paul took from one of his drawers something that he showed to his listeners. It was a rather heavy object, of a beautiful shiny yellow, and when exposed to the sun it gave out flashes of light. From its brilliance it might have passed for a metal.

"But that is gold!" cried Emile in astonishment at sight of the splendid stone; "a lump of gold as big as your two fists!"

"It is asses' gold, my boy," replied his uncle, "named thus by miners because it deceives the ignorant, and they take it for something precious, whereas it is really of small value. You can find as many of these stones as you please among the rocks in the mountains; but it wouldn't profit you a penny to pick them up. This substance is also called, in more learned language, iron pyrites, the last part of the term being taken from the Greek word for fire; for the stone will, in fact, emit sparks when struck with a piece of steel,—as, for example, with the back of a knife,—and these sparks are brighter than those obtained with flint and steel."

Here, by way of illustration, Uncle Paul made the stone that looked like gold give out brilliant sparks by striking it with a knife. Then he continued his talk:

"Iron pyrites or asses' gold has nothing about it of real gold but its luster and its yellow color. It is not a simple substance, but a compound of two elements that are familiar to you, though you would n't imagine them to be here, disguised as they are by the act of combination. One is iron, the other sulphur."

"That shiny yellow stuff that you would take for gold is made of iron and sulphur, like the ugly black powder in the artificial volcano?" was Emile's incredulous exclamation.

"It is made of iron and sulphur, and nothing else."

"But how different it is from either of them!"

"This difference comes from the sulphur's being repeated in asses' gold."

"The word, instead of making *ba,* makes *bba?*"

"Exactly. To indicate this repetition of the sulphur, they say in chemistry that iron pyrites is the *bisulphid* of iron, and you know that the first syllable of that term means twice."

"Then it 's the same as if they said 'twice sulphur and once iron.' The black powder in the artificial volcano is sulphid of iron, and this asses' gold is bisulphid of iron."

"Precisely. It could n't have been put better."

"I thank you, Uncle Paul, for showing us this splendid stone. It will make me remember that in chemistry *ba* and *bba* are not at all the same thing."

## CHAPTER VI

### EXPERIMENTS WITH THE BREATH

IMPRESSED by its glitter, the boys often talked to each other about the asses' gold which, despite its rich appearance, is made of just such sulphur and iron as go to make the black powder of the artificial volcano, but with a double quantity of sulphur. The magnificent stone left with them by their uncle they took delight in striking with steel, in some dark place, so as to produce bright flashes of sparks. Furthermore, directed by Uncle Paul, they resolved to visit some of the neighboring mountains in search of more stones like this one. So successful was their quest that Jules's cabinet became filled with pieces of iron pyrites of all sizes and of varying degrees of brilliancy. There were some golden yellow, cut in facets as if a lapidary had taken it into his head to polish them, while others were shapeless and more of an iron gray. Uncle Paul told them that the former were crystals, and that the majority of substances can under favorable conditions take regular shapes in which smooth facets arrange themselves according to geometrical laws. Such substances are then said to be crystallized.

"We will return to this subject later if opportunity occurs," said he; "but to-day other things demand our attention. So far we have been merely discussing, talking together, supporting our assertions by sundry facts picked up here and there. Your minds had to be prepared, had to become accustomed to certain ideas and expressions. But, now that you are ripe for it, we are going to have a little real chemistry; that is, we are going to perform some experiments. To see, touch, taste, handle, and smell for oneself, and to observe at leisure,—that is the only way to learn quickly and well. So, then, we will proceed with our experiments."

"Shall we have lots of them?" was the eager inquiry.

"As many as you please, my lads. Along that line, chemistry never comes to an end."

"Oh, that 'll be splendid! We shall never get tired of experiments. And may we repeat them by ourselves just as we did with the artificial volcano? That will make twice as much fun."

"If they are not dangerous there is no reason why you should not perform the experiments yourselves. When there is any danger I will tell you beforehand what precautions are necessary. I count on Jules to take the lead, for I know how careful and how skilful he is."

At this word of praise a slight color flushed the older boy's pale cheeks.

"Now, what shall we begin with?" said Uncle Paul. "It shall be with a substance that plays a most important part, air. Let me tell you at the outset, if you do not already know it, that air forms around the earth an envelop known as the atmosphere and having a thickness of about fifteen leagues at the most moderate estimate. It is a substance of an extremely subtle nature, so intangible and invisible that one is at first surprised to hear it spoken of as matter. 'What!' we exclaim; 'air is matter? Air has weight?' Yes, my boys, air is matter and can be weighed. With its delicate instruments physics can weigh air, and it teaches us that a liter of this invisible matter weighs one and three tenths grams. That is very little when compared with the weight of lead, it is true; but it is a good deal when compared with other substances that we shall soon learn about."

"Are there things lighter than air?" asked Jules in surprise. "Yet people say, 'as light as air,' as if there were nothing else of so little weight."

"Let them say it, but rest assured there are other things that in respect to weight are to air what wood is to lead. Air is colorless and, for that reason, invisible. Understand me correctly, however: when I say 'colorless' and 'invisible' I am speaking of air in small quantities; in large volumes that would no longer be true. Water will help us to understand this. Seen in a drinking-glass or in a bottle, it is colorless; seen in a deep body, as in a lake or the sea, it shows its blue color according to the depth of the water. Likewise with air: it is of a blue tinge, but so pale that to become perceptible the body of air must have enormous thickness. That explains why the sky is blue: the thickness of the atmospheric envelop (some fifteen leagues, as I said before) brings out its true color, which is imperceptible to the eye in a layer of moderate depth.

"Invisible, subtle, intangible, escaping the clutch of the fingers, air seems to present insurmountable difficulties to any one wishing to study it closely. If we desire to submit it to tests that will reveal its nature and properties, we must take a certain quantity of it, isolate it from the rest of the atmosphere, shut it up in some sort of container, make it flow out in this direction or that as we may choose, carry it from place to place, expose it to such and such conditions,—in short, make it obedient to our control as we should a piece of stone or a pebble. But how can we see the invisible, grasp the elusive, handle the intangible? The difficulty, you see, is no small one."

"It seems to me so great," replied Jules, "that I can't begin to guess how it is overcome. But I have too much confidence in you, Uncle, to doubt that we shall manage it somehow."

"We must; otherwise we should be held up at the very start. And that would be a pity, for air would not be the only thing to get the better of us. There are many other substances just as invisible, just as subtle, just as intangible as air, and of inestimable importance. They would all remain unknown to us if our present difficulty could not be overcome; and the great science of our day, chemistry, the mother of industrial wonders, would remain to be discovered in some ever-retreating future when the art of handling the intangible should have been mastered. All these substances, having the subtle and elusive quality of air, are known by the general name of gases. Air itself is a gas."

"And there is the gas they use for lighting, too," said Emile. "I thought that was the only thing called gas."

"What is burned in chandeliers in our cities is a gas, but not the only one. There are many others, each with its own peculiar qualities. The word gas, then, is a general term by which we designate all substances having a tenuity or thinness similar to that of air; and if we commonly restrict it to illuminating gas, it is because the latter is much better known to us all than any of the others except air. In ordinary language a general term is thus monopolized by one particular substance.

"But to return to our problem, how can we handle air, how subject any gas whatever to observation? I will show you. Suppose we wish to collect the air that comes from our lungs,—our breath, in short. I dip a tumbler into a bowl of water and fill it, after which I invert it in the bowl and raise it. As long as the brim remains completely submerged the water does not run out, but is held suspended above the general level of the liquid in the bowl. I see by your looks that this lifting of water and holding it motionless above the level of the surrounding liquid excites some surprise. I will return to this in a moment and explain the cause; but just now let us proceed with our experiment. Here is the glass, full of water and held up by one hand, with the brim immersed. Now with a glass tube—or, if necessary, with a reed or a big straw—I blow under the glass, and the air from my lungs makes the water bubble. Because of its superior lightness it makes its way upward in big globules through the contents of the glass until it reaches the inverted tumbler's bottom. As the breath—or, to express it better, the exhaled air—collects in the upper part of the glass, the water thus displaced descends and reënters the bowl. The thing is done: I have collected my breath; there it

is in the glass ready to undergo any tests we choose to apply."

"How easy it is, after all!" exclaimed Emile, much impressed by what he had just seen.

"It is nearly always so, my child,—very easy when we know how, very difficult when we do not."

"Then this glass holds what we send out of our mouth when we blow out a candle; I mean, it is filled with breath. It certainly is a curious thing to collect like this what can't be seen or felt. When I let out my breath after puffing up my cheeks, I don't see anything at all; and yet I just now saw your breath going up through the water and making it bubble."

"The commotion in the water made it seem to you as if you saw what is by its very nature invisible."

"Now that the water is still again, I see nothing, though I am sure the part of the glass that looks empty really has something in it; for I saw that something come and take the place of the water, which went down slowly in the glass. All the same, it seems to me very funny to have that glass full of Uncle Paul's breath. May I try filling it with mine?"

"Certainly; but first you must empty out what is now in the glass."

"Empty it out? But how?"

"In this way."

So saying, Uncle Paul took the glass and inclined it just enough to let a part of the brim come to the surface of the water, whereupon something escaped with a bubbling sound.

"It 's gone," cried Emile. "Let who wants to, run after it and catch it in the air where it has disappeared."

The glass being refilled with water, Emile took the straw and blew as his uncle had done, controlling the muscles of his cheeks so that he might watch the bubbles as they rose one by one; and great was his delight to see so easily, and to shut up in a glass so securely, what he had thought must always remain invisible and unmanageable.

"That was soon done," said he, when the glass was full. "I could fill a big bottle with my breath just as easily as a tumbler. May I, Uncle Paul?"

"You may, my boy. If you enjoy the experiment of bottling up your breath, I for my part enjoy your enthusiasm."

A large bottle of clear glass with a wide neck was standing on the table, having been placed there by Uncle Paul for later experiments. Emile took it up and went to the bowl, but soon saw that the latter was not deep enough to dip the bottle into so as to fill it and then admit of its being lifted up with the mouth under water as had just been done with the glass. "Look," said he, after a few vain attempts; "the way you did it with the tumbler won't work. What shall I do?"

"Since the difficulty does not yield to a frontal attack, let us turn its flank. Watch me."

Therewith Uncle Paul placed the bottle on the table and filled it from the carafe. Then, clapping the palm of his left hand over the mouth as a stopper, he took the bottle in his other hand, turned it upside down, and plunged it while stopped in this manner into the bowl of water. Then he removed his left hand, and the bottle with its neck immersed retained its liquid contents suspended above the exterior level, without losing a drop.

"You always find a way out, Uncle Paul," said Emile, delighted at this easy way of overcoming the difficulty.

"We must practise a little ingenuity, my child; for if we did n't, what could we accomplish with the poor apparatus our small village provides? Skill must make up for the defects of our appliances."

Emile blew, filling the bottle with his breath in a

few minutes. Then, after Jules also had performed the operation, so as to accustom himself a little to handling gas, their uncle continued thus:

"Why does the water in the glass and in the bottle remain above the level of that in the bowl? This is what we must now find out, though not in all its details, for to do so would take us out of chemistry into physics. A brief explanation, just enough to show you the cause of what now excites your surprise, is all that I at present propose.

"Air, I told you, can be weighed just the same as any other substance; and its weight, as I said before, has been found to be one gram and three decigrams a liter. That is very little, but the atmosphere is at least fifteen leagues thick, which must make an enormous number of liters piled one on top of another. Since, then, the atmosphere has weight it must press with all its force on objects immersed in it; it must press on them from above, from below, from the right, from the left, from every direction. It presses, for instance, on the water in our bowl; and the pressure, being transmitted by the liquid to the mouth of the bottle, keeps the water in the latter suspended above the exterior level.

"A striking experiment will convince you of this thrust exerted by the atmosphere. Over the mouth of a bottle filled with water we place a piece of damp paper, and while this is held in position with one hand the bottle is turned upside down with the other. Then the hand holding the paper can be withdrawn without the escape of a drop of water from the inverted bottle. It is the atmosphere pushing in every direction, upward as well as downward, that holds the water in. The office of the paper is to keep the air from entering the liquid mass and breaking it up, which would immediately cause the escape of the water."

"And shall we try this wonderful experiment?" asked the boys, in eager curiosity.

"Shall we try it? you say. Do you think I should have told you about it if we were not going to try it? Up and at it then! Here is our bottle, and here are paper and water; nothing further is needed."

The bottle was filled to the brim and a piece of damp paper placed over its mouth. With his right hand Uncle Paul raised the bottle by its bottom, holding the fingers of his left hand meanwhile on the paper. Then he carefully turned the bottle upside down, let go of the paper, and the thing was done: not a drop of water escaped from the bottle, even though its neck pointed downward. Emile, more excited than ever, could not contain himself.

"That 's fine!" he declared, "not a drop comes out of the bottle, and it 's turned upside down, too. If it had a cork in it, the thing would be natural enough; but the paper does n't cork the bottle; if you blew on it, it would come away. How long will the water stay in like that?"

"As long as you please; as long as one has patience to hold the bottle as I am holding it now."

"But the water is trying all the time to get out? It presses down and would fall if it could?"

"Yes, it keeps pressing down, and tends to fall, but the stronger pressure of the atmosphere restrains it."

"And what if we took away the piece of paper?"

"Immediately the water would run out, just as we have so often seen it do from a bottle or carafe tipped sidewise, or, still more, from one turned upside down. This piece of paper closes the bottle's mouth so that water and air are placed in a position where one pushes squarely against the other. Without it the water would slip through the air, the air through the water, and in this mutual evasion the bottle would speedily empty itself. Put two iron rods together, end to end, and exert pressure; there will be mutual resistance. That is what happens when we put the piece of paper between the air and the water. But if the iron rods were made into two bundles of very fine needles pushing against each other, end to end, these same rods would slip into and through each other just as happens with the air and water when there is no paper to separate them.

"To return to the bottle that Emile used to hold the breath he blew into it: as long as its mouth is immersed in the bowl the water it contains does not run out, but is held above the surrounding level by the force of the air pushing against it. Now, what would happen if instead of this bottle we used a very tall container,—a tube, for example, closed at its upper end? Would this tube, whatever its length, still remain full when raised out of the water except its lower end? No. If the tube were raised so as to project only ten meters above the surface of the water, it would indeed remain full; but if it projected beyond this height the part of the tube above ten meters would be empty. The pressure of the atmosphere can hold up a column of water only ten meters high; that is the extreme limit. Our containers here, as you see, are well within this limit in their height. However big or tall our bottles may be, there is no danger of their being so tall that the pressure of the atmosphere cannot hold back the water that fills them.

"Finally, suppose we wish to transfer a quantityof gas from one vessel to another, or to tranfuse it as they say. This gas shall, again, be our breath, which will perfectly well serve the purpose of the demonstration I have in view. I fill a glass with it by blowing through a tube in the manner just shown; and now I propose to make this volume of gas pass into another container, or it may be I wish to transfer only half of it. I fill this second glass with water and invert it in the bowl so as to keep only the brim immersed. The first glass, with only its brim still in the water, is then tilted sidewise under the other, whereupon the air it contains escapes in bubbles and passes into the second glass, wholly or in part, as I choose.

"To decant a liquid—to pour wine, for example,

from one bottle into another—a funnel is used, as you know. The same utensil is often very useful for decanting gases; but a chemist's funnel, which is likely to come into contact with all sorts of corrosive liquids, is of glass, a very resistant substance. As long as only gas is to be transfused, it will suffice to add a modest tin funnel to our simple outfit; but if we had a glass funnel it would be better and more in keeping with chemical practice. Furthermore, glass has one inestimable advantage over tin: it is transparent, and thus allows us to see all that takes place within it. But with nothing beyond a common tin funnel we are not necessarily brought to a halt in our operations.

"A funnel of some sort is indispensable for transfusing a gas from a container of any kind to a bottle with a narrow neck such as bottles commonly have. Of course the transfusion is effected under water. The bottle, filled with water and held with its mouth immersed, has the funnel inserted into it by one hand operating under water. That done, the jar or whatever it may be that contains the gas is brought under the funnel and inclined little by little until bubbles of the gas escape into the flaring mouth of the funnel and pass thence into the bottle.

"That will do for to-day, my boys. You are now in a position to repeat these experiments by yourselves as often as you like, collecting your breath in a glass, tranfusing it into another, or into a bottle turned upside down, thus getting your hand into practice. I shall soon need your assistance."

# CHAPTER VII

## EXPERIMENTS WITH AIR

UNCLE PAUL took a rather deep dish and in the center he fixed a candle with a few drops of melted wax. Then he lighted the candle and covered it with a large, wide-mouthed bottle of clear glass. After this he poured water into the dish until it was quite full.

Meanwhile the children looked on, chattering between themselves and wondering what their uncle was up to with his lighted candle all surrounded with water and burning in a bottle upside down in the middle of a dish. What curious experiment was he about to perform? They were not left long in doubt. Everything being ready, Uncle Paul began thus:

"What is there in the bottle?"

"A lighted candle," Emile hastened to answer.

"Is there nothing else there?"

"No, nothing. I don't see anything but the candle."

"You forget that there are some things we cannot

see. You must use here the eyes of the mind, not those of the body."

Emile scratched behind his ear with the tip of his finger and winked hard, as was his wont when puzzled. He was trying to think what the invisible thing could be that his uncle referred to. Jules came to his aid.

"It is air," declared the older boy. "There is air in the bottle where the candle is burning."

"But Uncle Paul did n't put any there," Emile rejoined.

"Is it necessary to put any there, little giddy-pate?" his uncle demanded. "Is not the bottle full of it without any help from us? All the vessels we use, all our flasks, jars, bottles, glasses, containers of whatever kind, being immersed in the atmosphere, in the very depths of this ocean of air, are filled with this gas as they would be with water if they were plunged uncorked into that liquid. If it does not contain anything else, every bottle, whether right side up or wrong side up, is full of air, which gets in of itself, usually without our paying any heed to it. When a bottle of wine is drained to the last drop, we say it is empty. Should we really call it empty if we chose our words with extreme care? Certainly not, for the so-called empty bottle is as full as ever, full to the top; it is filled with the air that has taken the wine's place. And so it is with all vessels when we empty them of their contents: if they were full before, they remain full afterward, though the nature of their contents has changed. Nothing is absolutely empty, nothing can be empty, when air is free to enter it. We can, it is true, empty a vessel in the literal sense; we can create a void or vacuum, as the learned men say; but that is an elaborate operation, and its success demands proper appliances."

"You mean an air-pump?" asked Jules.

"Yes, my boy, an air-pump, a special kind of pump that sucks the air from a tightly closed vessel and discharges it into the outer atmosphere. But as nothing of the sort has been used here, my bottle is full of air exactly like the air around us. Consequently the candle is burning in the midst of the air contained in the bottle. Now, why did I fill the dish with water? For this reason: The air in the bottle is the substance I propose to study with you by subjecting it to certain experiments, particularly this experiment with the lighted candle. And so we must separate this air from the rest of the atmosphere, isolating it in a closed container, as otherwise our experiment could not be carried through; and, furthermore, we should be unable to tell what portion of the atmosphere we had been experimenting with. The bottle alone does not furnish complete isolation, for, between its mouth and the bottom of the dish, air could easily slip in and mix with that already in the bottle. A barrier must be interposed to separate the inner from the outer air, and this barrier is furnished by the bed of water in the dish. Thus we obtain perfect isolation; and you will also see that the water serves a second purpose, being an indicator of what takes place in the bottle.

But I must not distract your attention by too much explanation. Now watch closely what is going on in the bottle."

Behold, in a few minutes the candle-flame, which at first was full and bright like that of a candle burning in the open air instead of in a bottle, began to get dimmer little by little, becoming shorter, shrinking in width, and looking dull and smoky. Soon it had diminished to a mere point, and finally to nothing at all. The flame was quite extinguished.

"Look!" cried Emile. "The candle has gone out without any one's blowing it out."

"Wait a moment, Emile, and we will presently talk the thing over. Just now keep your eyes open and watch what is happening in the water, the indicator I spoke about."

Emile and Jules watched attentively and saw the water gradually rise in the neck of the bottle, completely fill it, and go still farther, so that an appreciable part of the bottle, at first filled with air, was occupied by the liquid coming from the dish and slipping in from beneath. Its ascent was slow, and when at last it was finished Uncle Paul broke the silence.

"Now," said he, "you may ask all the questions you please."

"I should like to have one thing explained," said Emile. "When a candle is burning and you want to put it out, you have to blow on it. But here no one blew out the candle, and if we had wanted to we could n't have done it on account of the bottle over it. Not a breath of air came, that is certain; no puff of wind. Under cover of the bottle no breath of air or puff of wind could get at it. The flame was perfectly still; it stood up straight and calm; and yet, without any reason that I can see, it got dim and dwindled to nothing but a point, and then at last went out altogether."

"I should like to ask something, too," put in the older boy. "The bottle was at first filled full of air; now, besides the air that is left, it has several fingers of water from the dish. I saw this water rise little by little as the flame dwindled. So something has been taken from what was in the bottle at first, for the water rose and took its place. But how did that something disappear, and where has it gone to? If you had not told us and made us understand that nothing is ever annihilated, I should say that a part of the air was being annihilated while the candle burned."

"Jules's point shall have our attention first, as it will furnish us the explanation of Emile's difficulty. A part of the bottle's gaseous contents has disappeared, apparently; the water's rising above its level in the dish to fill the vacancy would seem to prove it in a way to convince the most incredulous. Something, I repeat, has disappeared, so far as we can see; but we must not for a moment suppose it has been annihilated. Let us investigate further, and we shall discover what has become of this something that seems to be lacking.

"I have already told you that heat and light are

nearly always sure signs of the mingling of substances of different natures,—of chemical combination, in short."

"I remember," said Jules. "You called it the fireworks that celebrate chemical weddings. Can such weddings take place in a bottle?"

"Yes. The flame was hot and it gave out light; hence a chemical combination was going on, producing this heat and light. And what substances were they that were thus combining? One of them, beyond a doubt, was supplied by the fatty material of the candle as it melted under the heat from the lighted wick; the other could have come only from the air, since the bottle contained nothing else. From this combination sprang something new, something that was no longer either candle grease or air, something with properties that neither air nor candle possesses. The compound thus formed is an invisible substance, a gas just as air is a gas, and for that reason invisible."

"But if a gas is made out of air and candle grease," Jules objected, "this gas takes the place of the air that disappeared, and the bottle ought to be just as full as it was before. I don't see why the water rises."

"Wait; we are coming to that. The compound we are speaking of is readily soluble in water, somewhat as sugar and salt easily dissolve in that liquid. Once dissolved in this way, sugar and salt disappear, become invisible, and the only evidence we have of their presence is the sweet or the salt taste of the water. In like manner, what is produced by the flame disappears: it enters the water and becomes incorporated with it. Similar but much richer gaseous solutions are familiar to you. I hardly need to remind you of beer, cider, sparkling wine, and soda-water, drinks that make the cork pop and that foam when poured into the glass. All these liquids hold quantities of gas in solution, —so much of it, in fact, that they cannot retain it when escape is possible; and therefore it thrusts out the cork and covers the drink with foam. Now, curiously enough, the gas that makes these beverages foam is precisely the gas that the candle-flame manufactures. Some day I will take this interesting subject up again. I mention it here in passing, but have not time to dwell on it.

"Since the compound produced by the candle grease and the air disappears from sight, being dissolved in the water, a vacant space must be left; and this it is that the water in the dish, under pressure of the atmosphere, rises in the bottle to occupy, showing us by the height to which it rises the amount of air that has disappeared."

"It has n't risen far," said Emile. "See! It 's only just above the neck."

"That shows that the candle burned up only a small part of the air in the bottle,—a tenth part, let us say, if the water that has made its way in fills a tenth part of the bottle's total capacity."

"But as there is still so much air in the bottle, why did n't the candle use that up just as it did the other part? I don't see that it is any different from what that was. It is still transparent, invisible, and has n't a particle of smoke."

"Here is where we come to your question, my boy, the question why the candle went out without being blown out. The candle-flame is caused by the combination of the material of the candle with certain other material contained in the air. Air and candle are equally necessary in feeding the flame. If either be lacking, the flame is extinguished. As to the candle's being necessary, that is plain enough: no fuel, no flame. But in regard to the air you don't feel quite so sure. And yet what you have just witnessed ought to give you food for reflection. If the candle went out of its own accord, there was certainly something wanting."

"I 'm ready enough to admit that. Yes, that must have been the trouble; something was wanting, as no one blew out the candle and there was n't a breath of wind. Now, what was it?"

"What was lacking must have been air, as the bottle contained nothing else to begin with. Air is indispensable for keeping a flame burning."

"But there is still air in the bottle,—lots of it, almost the whole bottleful."

"I do not deny it; but listen a moment. Is it not possible that air, instead of being made of only one substance, is made of two of different kinds, equally colorless, equally subtle and invisible, and very thoroughly mixed together? Is it not possible, too, that one of these gaseous substances can keep a flame going, whereas the other cannot, and that the first-named forms but a small part of the atmosphere as compared with the second? If so, when this first part is used up in the bottle, the candle goes out of itself, as we say, the candle grease no longer finding what it needs for burning, and there will be left in the bottle, colorless and invisible, as it was in the first place, the other gaseous element, of no use for keeping a flame alive."

"How clear that makes everything, Uncle," said Jules. "Now I see it all perfectly. The candle went out of itself when there was no more of this gas material to keep it burning. By combining with a little burnt grease, this matter turned into something else, a gas that in dissolving disappeared in the water while the water in the dish rose and took its place. Now the bottle holds only the kind of gas of no use to a flame, and that is why the candle stopped burning."

"Yes, that 's the way of it, with one slight correction. A candle-flame is not by any means a fire strong enough to use up all the gaseous element needed for the act of burning; there is still some of it left over, but too little to keep the candle alight. The air has become impoverished, but not deprived of all it contained of the element in question. Another day we will try to find a way to remove the last trace of this flame-supporting gas. At present let us rest content with our partial result. The contents of the bottle will no longer keep a candle burning, and a lighted candle put into the bottle would immediately be extinguished."

"A candle would go out if put into that bottle?" asked Emile, still rather inclined to doubt.

"Certainly it would, and almost as quickly as if you plunged it into water. How can you expect it to burn? If the one that was there could not burn, why should another do any better? They are all made alike."

"All the same, I 'd like to see it tried."

"Your curiosity shall be gratified."

With this Uncle Paul took a short candle-end and tied it to a wire bent up at the bottom. Then, raising the bottle a little with one hand, he slipped the palm of the other over the mouth, which was still under water, thus stopping it up. After this he set the bottle upright on the table without letting it lose any of its contents, liquid or gaseous, and withdrew the hand he had used as a stopper.

"But the air in the bottle will get out," objected Emile, "if you leave the bottle open."

"There is no danger of that," his uncle reassured him. "The invisible gas there, being as heavy as air, will not escape. However, to make sure of it, here is a stopper we can use."

It was a little piece of glass from a broken window-pane. Uncle Paul placed it over the bottle's mouth.

"Now," said he, "let us proceed to our experiment."

The candle fastened to the wire was lighted and allowed to burn until the flame was bright and full, after which Uncle Paul removed the piece of glass and gently let down into the bottle the lighted candle, which almost immediately became dim and went out. A second attempt met with the same speedy result. If violently blown upon or plunged into water, the candle could not have gone out more promptly.

"Well, are you convinced now, Master Unbeliever? Here, try it for yourself, so as to be quite satisfied."

Emile took the candle and began the experiment over again, lowering the flame little by little, very gently, very cautiously, so as to keep it away from the sides of the bottle, thinking that careful management might accustom the flame to this unsuitable atmosphere. But it was of no use. Though several times repeated, the attempt invariably failed: a little higher or a little lower in the bottle, the flame always went out.

"It 's no good, the candle won't burn there," declared the boy, tired of attempting the impossible. "I should be perfectly sure of it if I were certain the bottle had nothing to do with it. Could n't the nearness of the glass and the want of space make the candle go out?"

"That is a very natural question, but it is soon answered. Here is a bottle like the first, of the same size and with the same wide neck. It is filled with air that has not been impoverished by anything burning in it, the same kind of air as that all around us. Repeat your experiment with that."

Emile lowered the candle into the bottle, and it burned very well, exactly as if in the open air. Whether put in abruptly or gently, near to or away from the enclosing glass, it remained alight and burned as it had outside. His constant failure with the first bottle and his repeated success with the second dispelled Emile's last lingering doubt.

"I have nothing more to say," he declared. "The stuff in the bottle where the first candle burned does n't suit it now at all."

"You are convinced, then?"

"Yes."

"Completely?"

"Yes."

"Then I will continue. What we conclude from our experiment is that air is composed of two different gases, equally invisible but of so unlike a nature, each to the other, as to prevent our confusing them. One, the less abundant, suits the candle-flame and combustion in general; the other, more plentiful, does not. The first is called oxygen, the second nitrogen. They are two simple substances, two metalloids. As to air, which is a mixture of these two gases, we can no longer properly call it an element as did the ancients. It is a compound of two substances of very different natures. This has been known only a comparatively short time."

"I wonder why it took so long to find out about air," said Jules. "It would have been such a simple thing to burn a candle in a bottle turned upside down in a dish of water."

"Very simple, undoubtedly; but it had to be thought of, and there was the difficulty."

## CHAPTER VIII

### FURTHER EXPERIMENTS WITH AIR

"OUR experiment with the candle burning in a bottle turned upside down in water, is one of the easiest to perform, calling for nothing that is not readily procurable. But, unfortunately, it is an incomplete experiment, for the reason I have already indicated. It shows us that air is composed of two different gases, one that will keep a flame burning and is called oxygen, and another of a contrary nature in this respect and known as nitrogen; but it does not tell us how much there is of each, because what is left after the candle goes out, still retains a good proportion of oxygen instead of being reduced to pure nitrogen.

"A candle-flame is delicate; a moderate puff of wind will blow it out. In the bottle, I admit, it is sheltered from any current of air, but its weakness prevents its using up all the gas on which it could feed. It turns dim and then goes out altogether when this gas begins to get scarce. If the comparison is not too far-fetched, one might call the candle-flame a guest at table with a poor appetite, leaving on his plate the greater part of the food served to him. Let us, then, call in a guest with a stronger stomach and able to eat his food to the last morsel, leaving nothing but the uneatable, the bare bones. I mean, let us find a kind of fuel that burns with enough energy to consume all the oxygen to the last trace and leave only the useless gas, the nitrogen.

"What shall this fuel be? Shall it be coal? No; that would not do any better than candle-grease, burning less freely, in fact, because needing the heat of a glowing furnace to keep it going, and that is out of the question in our experiment, which owes more than half of its value to its very simplicity. Shall it be sulphur? There, assuredly, we have a strong stomach needing no second invitation when oxygen is offered. Once set fire to, it burns vigorously. But it has its faults,—its suffocating fumes. Nevertheless, I should be glad to avail myself of it if there were not something better at hand. You are familiar enough with the common match, the little stick of wood tipped with sulphur, and lightly coating the sulphur we find— See who can tell me first."

"Phosphorus!" cried the two listeners at the same time.

"Yes, phosphorus, which is inflammable to a degree not attained by any other substance in common use; phosphorus, which takes fire by being merely rubbed against the sandpaper cover of the match-box or against a rough wall. Nothing equals it in the vigor and persistence with which it burns. Here, truly, we have the greedy guest who will leave nothing on his plate. But first let us get a little better acquainted with him. Phosphorus is not very well known to you, as hitherto you have seen it only on the tips of matches."

"Sometimes," said Emile, "the tips are red, and sometimes blue, or yellow, or almost black. Does phosphorus have all those colors?"

"No; of itself phosphorus has only one color, which is nearly that of yellow wax. But the match-manufacturer adds colored powder, sometimes of one hue, sometimes of another, according to his fancy, for the sake of giving a little variety to his wares and thus pleasing the purchaser's eye. Glue is also mixed with the phosphorus to make it stick to the sulphur. So what you are used to seeing is not pure phosphorus; but I will show you some that is perfectly pure.

"A few days ago, being called to town on business, I bought a number of things that our laboratory was much in need of. A laboratory, let me explain to you, is a place devoted to scientific research; it is the scientist's workshop. Modest though our workshop is, it must have some equipment, certain implements and supplies; otherwise we should have nothing but our ten fingers for an outfit, and what could we do with them? We should simply have to content ourselves with talking and nothing else. But that I will not have, for I do not think much of chemistry carried on in words only. I wish to give you facts, things that you can see over and over again, substances that you can feel of, taste, examine, handle for yourselves, as that is the only true way to learn.

"What could the blacksmith do without his anvil and hammer? Nothing. Equally helpless is the chemist without the various appliances and drugs of his laboratory. We will furnish ours, then, little by little, but in a very modest way, I assure you beforehand, as your uncle's resources do not permit luxury. We will have the indispensable, but nothing more. Nor is it altogether a misfortune to be thus forced to use one's wits a little, in devising ways and means to make what one has suffice, and in getting along without what one does not have. Our earthen dish borrowed from the kitchen, our old medicine-bottles and preserve-jars—did they not play their part well? I assure you in all sincerity, we could not have done better if we had had the outfit of a costly laboratory. Why should n't we continue our studies in this way, as far as may be? If you ever chance to have access to a real laboratory, and to work in it, my little lads, you will take pleasure in recalling your uncle's poor outfit and in reflecting how little it took to lay the solid foundations of useful knowledge in your minds, and how little it would need to do so for others even in our smallest villages.

"It may well happen that we shall be halted now and then by difficulties impossible for us to overcome; then, and only then, we shall be forced to appeal to the expert chemist for aid. Such is our position to-day. We had need of phosphorus, and here it is, bought recently at the druggist's in town."

Uncle Paul here set before his nephews a bottle holding in water a yellow substance in the form of a stick as long and as thick as one's little finger.

"That," said he, "is pure phosphorus. It is semi-transparent, which, with its color, makes it look like a pretty piece of wax such as we have in the honeycomb. It is like wax before being bleached at the taper factory by long exposure to the sun."

"Why do you keep it in water?" asked Jules.

"I keep it in water because if exposed to the air it would soon catch fire. So inflammable is it that the slightest heat is enough to ignite it."

"But the phosphorus on matches does n't catch fire like that; you have to scratch it."

"I told you that in matches it is not used in a pure state. It is mixed with certain other substances with glue and colored powder, which lessen its inflammability. But even so it sometimes occurs in summer that matches take fire of their own accord. This is a serious fault which, added to others, will perhaps some day induce us to give up the use of phosphorus, when science has discovered something better to take its place."

"And why," asked Emile, "does n't it catch fire in the water, if it is so eager to burn?"

"Has Emile forgotten what I told him yesterday? In order to have fire, two things are necessary, each as indispensable as the other,—the thing that burns and the thing that makes it burn, this latter being a gas called oxygen, contained in the air. Combustion is brought about when the two combine. Where there is no air—or, rather, where there is no oxygen—combustion is impossible, however inflammable the fuel may be. So I guard the phosphorus from risk of taking fire by keeping it in water, which protects it from the air. This being excluded, we are sure the phosphorus will not catch fire.

"Still another precaution is not out of place with this dangerous substance. The bottle holding the phosphorus in water might get broken, which would expose the phosphorus to the air. Therefore we must guard against any chance shock or fall, and to this end the bottle is enclosed in a tin box which serves as a shock-absorber. In this double enclosure and immersed in water, phosphorus is kept indefinitely on the druggist's shelf without any danger.

"It remains for me to add that a burn from phosphorus is a serious accident, a most painful accident. There is nothing that smarts more than a wound made by this terrible stuff. Neither live coals nor red-hot iron can cause so acute and lasting a pain. I will leave you to imagine the fate of the thoughtless person who, wrapping a piece of phosphorus in paper, should put it into his pocket, proposing to amuse himself with it later and make it shine in the dark. The heat of his body would set fire to his perilous possession, and the imprudent one would be burned to his very entrails, while he filled the air with his shrieks of agony. Be careful then, my boys, not to play with this terrible substance. If the thirst for knowledge should tempt you to handle it at all, do so with the utmost caution.

I appeal here to Emile's obedience and Jules's prudence. Had I not perfect confidence in you, did I not know you to be incapable of reckless folly, I should double-lock and triple-lock my arsenal of drugs and banish phosphorus from our lessons forever.

"I should be all the more impelled to this course because the risk of fire and serious burns is not the only danger to be considered; there is another peril, which you must be still more careful to guard against. Phosphorus is a deadly poison, a few particles of it being enough to cause death with frightful sufferings. I will say no more; you are warned. Look on phosphorus as one of the most formidable of foes, and let no carelessness on your part lay you open to its assaults.

"After these warnings, dictated by prudence, I will now explain how phosphorus may be used to show what air is made of. We must burn a little in a certain volume of air properly separated from the rest of the atmosphere. Our container in this operation should be of considerable size, so that the glass may be far enough from the flame to escape the risk of being cracked or broken by the heat. A common glass preserve-jar holding two liters or more, and as large at the top as at the bottom, would do very well if I had not something better. This something better is a chemist's bell-glass, a recent purchase of mine, which we shall find to be one of the most useful appliances in our laboratory. I will ask you to be especially careful of it. It is, as you see, a simple container of colorless glass, cylindrical in form, with a dome-shaped top surmounted by a little knob for taking hold of. Some are made with flaring mouth, reminding one of the shape of our bronze church bells, whence comes the name of bell-glass. For certain delicate plants requiring shelter and warmth gardeners use similar glass covers, but they are commonly too large and cumbrous for laboratory use. If one could find a medium-sized one, it would do perfectly.

"A large bottle, a gardener's glass plant-protector of suitable size, or the ordinary chemist's bell-glass—any one of these that should chance to fall into our hands would serve as a container for burning phosphorus; but Uncle Paul's purse has procured for us a real laboratory bell-glass, so let us be grateful for it and proceed with our experiment.

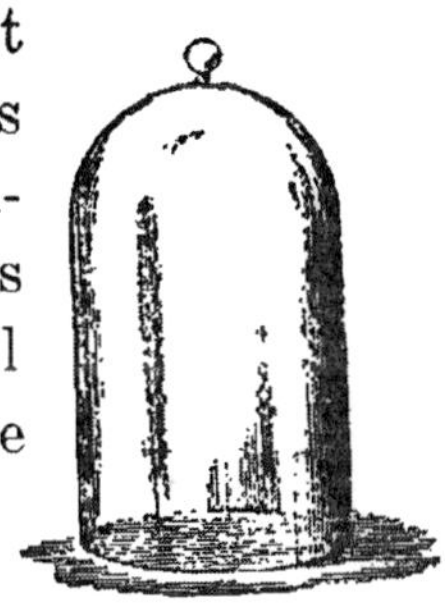

"The combustion, the burning of our bit of phosphorus, must take place on water, to prevent any communication between the air in the glass and that outside. Consequently, the phosphorus has to be placed on a tiny raft that will keep it dry, and for this purpose we can use any small object that will float, as a piece of cork or a bit of wood. But this float of ours would catch fire if unprotected from the burning phosphorus. Accordingly, we will put the latter in a tiny earthenware cup resting on the float; and for our cup we will simply take a concave fragment of some old broken pot. Now all is ready and we will proceed to business.

"First we must cut off a piece of our stick of phosphorus. It is soft enough to cut with a knife, being of about the hardness of rather firm wax: but it is not a thing to be cut as carelessly as one might whittle a stick of pine wood, the mere friction of the knife being likely to prove sufficient to set fire to the phosphorus if it were exposed to the air, with consequent serious injury to the clumsy operator. The inflammable stuff should not be held in the air except by the finger-tips and for as short a time as possible, and the cutting should take place under water. Watch me."

Uncle Paul put his fingers into the bottle and drew out the phosphorus stick, which gave out a rather strong smell of garlic, with some slight wisps of white smoke. The children were told that this smell of garlic is the natural odor of phosphorus, and that the white smoke would be found to give out light if looked at in the dark. Matches emit, to a lesser extent, these same odorous fumes. The phosphorus was immediately plunged into the bowl of water, and there, both hands under water, Uncle Paul with a knife cut off a piece about as large as two peas. This fragment was placed on a bit of broken crockery, and that on a little wooden raft of sufficient buoyancy to float its load; then the whole thing was set on the surface of the water in the middle of the bowl. A lighted match quickly started the phosphorus to burning, and Uncle Paul hastened to cover it with the bell-glass, which was, of course, full of air.

Behold, then, the phosphorus blazing away with

a violence quite new to the boys, who up to this time had seen no more of the inflammable stuff than the minute quantity at the tip of a match. The flame crackled, the light was brilliant, almost blinding. A dense cloud of white smoke formed, giving the appearance of milk to the contents of the bell-glass. At the same time the water in the bowl rose so rapidly in the glass that Uncle Paul was obliged to add more, in order not to leave the bottom of the bowl dry, for that would have let air into the bell-glass. So thick was the milky-looking cloud that the phosphorus flame could no longer be seen; or, if seen, it was only at intervals, like lightning in a mass of clouds. But the jets of light became more and more infrequent and feeble, and finally ceased altogether.

"It is over," Uncle Paul announced. "The phosphorus has used up all the oxygen in the air contained in the bell-glass, and there is nothing left but nitrogen, which will not support combustion, although there is still some combustible matter left on the bit of crockery. We shall see it when the white smoke has cleared away. Meanwhile let us talk a little about this smoke, which seems to attract your attention by reason of its beautiful milky appearance. It comes from burned phosphorus—that is to say, from the phosphorus combined with the oxygen of the atmosphere. A brilliant light, of such intensity as to try our eyes, accompanied this act of combining, as it always does. I say nothing of the heat, to which the bit of broken crockery could testify if it could speak. These fumes are easily dissolved in water, and thus there is left a vacant space which the water from the dish rises to fill, little by little, in this way showing how much oxygen has disappeared. We should have to wait only about twenty minutes, more or less, for the contents of the glass to become as clear and transparent as at the start. But to hasten the process and not to put your patience to such a test, let us see what this will do: we shake the bell-glass gently so that the moving water washes the interior and takes up the smoke. By this operation the contents will soon be made clear."

With a little careful management in shaking the glass, the gas within was soon made to resume its original transparency, and then there was revealed, on the bit of broken earthenware, the residue of what had been placed there, but now of a reddish color and, indeed, so changed in appearance that the boys would not have recognized it as phosphorus. Melted by the heat so that it was spread out on the piece of earthenware, it had quite an altered look. But to convince his hearers that it was still phosphorus, their uncle tilted the bell-glass slightly, so as to bring the little raft near its edge, when it became an easy matter to withdraw the raft and its load.

"What we have here," said he, "is really phosphorus despite the reddish tinge that heat and melting have given it. There is even more left over than there was burned. You shall judge for yourselves."

The potsherd was taken out into the garden so as not to mingle the disagreeable phosphorus fumes with the air of the workroom, then with a match the reddish substance was set fire to, and it burned with the bright light and dense white smoke attending the combustion under glass. Thus it was proved that there remained some phosphorus, even a good deal, as it burned for a considerable time; and in this instance every particle of it was consumed, the last trace being dissipated in the air as white smoke.

"If combustion stopped under the bell-glass," Uncle Paul continued, "it was not for lack of something to burn, for there was a good deal of it left at the end, but for lack of the gas necessary to support combustion—oxygen, in a word. It stopped when the last trace of this gas was used up, phosphorus being able to burn as long as there is any oxygen left, however little it may be. Consequently, the bell-glass now contains nothing but pure nitrogen, a gas in which no substance whatever can burn.

"The phosphorus experiment tells us once again, but more positively and more distinctly, what the candle experiment told us: the atmosphere contains two gases, oxygen, which supports combustion, and nitrogen, in which neither candle nor phosphorus nor anything else can burn. It tells us, also, in what proportions the two gases, both simple substances, both metalloids, are combined in the atmosphere. Our bell-glass is cylindrical. If we divide its height into five equal parts, these will represent equal capacities, equal volumes. Now we see that the water that rose in the glass and took the place of the departed oxygen has mounted to a fifth of the total height, nitrogen occupying the other four-fifths. Thus the air about us has four times as much nitrogen as oxygen; or, to express it differently, in five liters of air there are four of nitrogen and one of oxygen.

"We will stop here for to-day. To-morrow, let me notify you in advance, our chemical experiments will call for two live and uninjured sparrows. Set your snares and catch them. I must ask you also to be careful not to molest any of the various species of garden birds, industrious hunters of insects and worms that are the scourge of agriculture; but I gladly give you a free rein in regard to those pillaging sparrows that, eager to find some tender foliage to tickle their palates in the spring, fly down from the neighboring roofs and nip my peas as fast as they sprout. I must have two of these birds for our instruction and to serve as a lesson to their brother marauders."

## CHAPTER IX

### THE TWO SPARROWS

THE two sparrows had been caught. Spring-traps, hidden under a thin layer of earth amid the rows of sprouting peas, and each baited with a piece of bread, had soon accomplished their purpose. Extricated from their entanglement in time to prevent strangling, the two captives were now moving about, full of life, in a cage. The boys were impatient to know what their uncle proposed to do with them, eagerly looking forward to some highly interesting and important experiment. A course of study having for them the keenest interest, and carrying with it the fun of catching sparrows, was real play in their eyes,—a circumstance highly pleasing to their uncle, who was convinced that in order to learn well one must enjoy learning.

"Ever since yesterday," said he to them, "the bell-glass full of gas, in which the phosphorus would no longer burn, has been standing in the bowl of water. The white smoke left after the experiment, and of which a little might have continued to linger all the afternoon, has now had ample time to disappear, to become absorbed in the water, so that at present there is nothing in the glass but perfectly pure nitrogen. Notice the transparency, the complete invisibility, of this gas. Would n't you take it for ordinary air, air such as filled the bell-glass in the beginning? It looks just the same, and yet how different are its properties! In this gas nothing will burn, no matter what we try. After what has occurred, this is plain enough without further demonstration. Some phosphorus—a good deal of it, in fact—was left inside the bell-glass, as was proved when we burned this residue out in the garden. If this remnant of phosphorus could not burn under the glass, but did burn very readily in the open air, the reason must be that it no longer had around it the gas needed to support combustion, having consumed what was originally contained in the bell-glass. In the open air is found an unlimited supply of what it had lacked under the glass, and that is why it began to burn again more brightly than ever and continued to do so until the last particle of the phosphorus was used up.

"Hence it is plain enough that as phosphorus could not go on burning in the gas now remaining in the bell-glass, no other substance whatever can burn there. Where the most inflammable of substances stops burning how could one less inflammable remain on fire?"

"That is very clear," admitted Jules; "what the strongest cannot do, certainly the weakest cannot. Then this gas—this nitrogen, as you call it—would immediately quench any fire that could be plunged into it?"

"Certainly. No burning substance plunged into it could go on burning for a single moment."

"It would be the same as when the candle stopped burning in the bottle? Emile, with all his care, could n't make it stay alight."

"Yes, that 's it, except that it falls short of the truth. I told you a candle-flame has n't energy enough to use all the oxygen in the air. A considerable part is left over in any such experiment as we performed at first; that is to say, in the experiment with the candle burning inside the bottle turned upside down, what is left of the gas is not pure nitrogen; there is still a little oxygen mixed with it, but not enough to keep a second candle alight where the first one ceased to burn. We found, in fact, that we could n't put a lighted candle in there without its going out immediately. But a more inflammable substance, such as phosphorus, would find what it needed in the residue of oxygen there, and would go on burning for some time where the candle could not."

"One might, then, put it this way," suggested Jules: "phosphorus is so hungry for oxygen that it licks up the leavings of the candle, which has a more delicate appetite."

"Yes, that would express it admirably. If there are any remnants of oxygen, phosphorus, with its robust appetite, will be sure to devour them; but if there are no such leavings, it must go without, and in that case it ceases to burn just as any other combustible material would have to do."

"That seems clear enough," said Emile, in his turn; "but, all the same, I should like to see it proved by experiment."

"What you desire is exactly what I propose to do," replied his uncle; "only we must first transfer a little of the gas from the bell-glass to a wide-mouthed bottle, in which our tests can be made more conveniently. So now is the time for us to put into practice the little I have told you about the transfusion of gases. As our bowl is too small and too shallow for this purpose, we will use this large tub filled with water."

So saying, Uncle Paul lowered the bowl with the bell-glass into the tub without disarranging them, and as soon as the rim of the glass was immersed he withdrew the bowl. A wide-mouthed bottle full of water, inverted and with only its mouth immersed, was held in position by Jules. His uncle tilted the bell-glass a little, and made some of the gas pass into the bottle and fill it. Then the bowl was again slipped under the bell-glass, and the whole thing was replaced on the table. Finally, the bottle, filled with nitrogen and stopped with the palm of the hand, was set upright on the table and a piece of glass laid over its mouth. In these various operations, harder to describe than to execute, care was taken never to leave the containers of nitrogen open to the outer air, an indispensable precaution and one easy to observe by keeping their mouths immersed and operating under water.

"Here we have our bottle full of nitrogen," said Uncle Paul. "Now what shall we try first,—sulphur, phosphorus, or candle?"

"Let 's begin with the weakest," was Emile's suggestion, "and try the candle first."

The candle, tied to a wire, was lighted and slowly let down into the bottle. Scarcely had it passed the mouth when it suddenly went completely out, not retaining even for an instant that red glow on the wick which is wont to linger for some time after the candle-flame has been blown out. A plunge into water could not have caused a more instantaneous and complete extinction.

"Ha!" cried Emile, "that goes much better than when we tried it before. Yesterday the flame seemed to hesitate, sometimes, about going out; it had to be let down well into the bottle, and the wick appeared to want to keep its red spot; but to-day we see nothing of the sort. As soon as the candle was lowered into the neck of the bottle, flame and red spot both vanished at the same time. And now let 's try phosphorus."

"Phosphorus will not burn there any better; you shall see."

The fragment of broken crockery again served as a cup the size of a five-franc piece. An iron wire was bent at one end into a ring for holding this cup, which in turn was to hold the piece of phosphorus. These arrangements completed, the phosphorus was first lighted, and then lowered by means of the iron wire into the bottle of nitrogen. It went out abruptly, as had been expected. What burned so furiously outside the bottle immediately refused to burn when put inside.

A similar test was made with sulphur, which Emile seemed to think might burn in the bottle because of its high inflammability; but it went out as quickly as the candle and the phosphorus.

"Further tests are needless," declared Uncle Paul; "they would only lead to the same result. Nothing can burn in nitrogen; or, in other words, that gas will not support combustion.

"So now we will proceed to make use of your two sparrows, whose part in our study of chemistry is still a puzzle to you. They will teach us some very interesting things, to atone in part for our loss in the peas they have destroyed. In the first place, we must fill our bottle once more with nitrogen. What is still there has come into contact with phosphorus, sulphur, and candle-grease; therefore we are not certain that it is perfectly pure, as it should be for the purpose of our experiment. Accordingly, we will obtain a fresh supply from the bell-glass, first emptying out what is already in the bottle. But how shall we proceed to accomplish this?"

"To empty a bottle, all you have to do is to turn it upside down," Emile made haste to explain, without stopping to think.

"Yes, if the bottle contained water or some other liquid," replied his uncle; "but it contains a gas of about the same weight as air. If you should try to empty a bottle of air by turning it upside down, you would never succeed."

"That 's so. Then let 's blow as hard as we can into the bottle, and we shall drive out what is in it."

"Agreed; but first tell me how we are to know when we have driven it all out, there being nothing to show what goes out and what comes in. Furthermore, you will only replace the first gas with another, your breath, which will be just as hard to drive out; and so the thing will have to be done all over again, and after that still again, and so on without end."

"Really, the more I think of it, the harder it looks to me. I was a little too hasty when I said it was so easy to do. Jules says never a word, and I 'll bet he does n't know any better than I how to manage it."

"I own I am puzzled," said Jules. "This little matter, which seemed a mere nothing, brings me up short."

"It will not stop you very long. Here is how it is done."

His uncle took the bottle and plunged it into the tub, where it quickly filled with water.

"There you have the gas driven out completely."

"Yes," agreed the boys, "but now the bottle is full of water."

"And what is to prevent our replacing it with nitrogen from the bell-glass just as we did before?"

"Why, that 's so! It 's as easy as can be. The only hard part was to think of it in the first place, just as you said yesterday."

"I am reminded at this point," said Uncle Paul, "of something that may appropriately be mentioned here. In order to ascertain whether the composition of the air is everywhere the same, aëronauts and travelers sometimes bring back air taken at the height they have reached. Now, how can a sample of air be taken at the summit of Mont Blanc, for instance, or in the lofty altitudes attained by balloonists? How make sure that the air really comes from precisely such and such a place, from a given mountain-top, for example, or from a certain height in the heavens above? Imagine a series of bottles labeled: 'This is from the top of Mount Perdu,' 'This was taken at a height of eight thousand meters, by an aëronaut,' 'This was brought by vessel from such and such a latitude and longitude at sea.' How are specimens to be obtained from various distant points when they are desired for the purpose of chemical examination? Nothing easier. A bottle filled with water is emptied at the exact spot where it is proposed to take a sample of the air, and the air at this spot rushes in to take the place of the liquid poured out. Then the bottle, carefully corked, will hold henceforth, with no further precautions, the invisible substance that at first seemed so hard to collect in perfect purity.

"We now come to the sparrows, whose part in all this I see you are impatient to learn. From the contents of the bell-glass I once more fill our bottle with nitrogen in the manner already shown. A second bottle of the same size and shape, but full of air, is placed on the table beside the first. There they both are, with pieces of glass over their mouths as stoppers. In the appearance of their contents there is no difference, each bottle showing the same clearness, the same invisibility, as to what is inside. Now I will put these two sparrows into our bottles, which are large enough to hold them for the short time required by the experiment. But first I will ask Emile which bottle, if he were a bird, he would prefer to go into, the one with air or the one with nitrogen."

"A week ago," replied the boy, "I should have said it didn't matter which I chose, for there is nothing to be seen in the one any more than in the other; but now, to tell the truth, I 'm beginning to be afraid of these invisible things. That rascally nitrogen that puts out a candle is not to be trusted. I don't know much about it, and I do know a little more about air, and so I 'd rather trust air than nitrogen. If I were a sparrow, then, I 'd choose the bottle filled with air."

"And you would choose very wisely, as you shall soon see."

Taken from the temporary cage, the sparrows were put, one into the bottle full of air, the other into the bottle of nitrogen. The piece of glass laid over the mouth of each bottle shut in its occupant completely. The young observers looked on, deeply interested in what was to come next. In the bottle of air nothing unusual occurred. The captive fluttered about, pecking at its prison-wall of glass, that mysterious obstruction which it could not see and yet could not pass through. It tried to take flight, fell back, rose again, and recommenced its vain attempts. All this was simply the agitation of a bird seeking to regain its lost liberty, a desperate attempt to escape from the prison, and nothing more. Vigorously alive, struggling with beak, claw, and wing, the bird evidently had no other feeling than one of extreme fright.

Quite otherwise did the sparrow in the bottle filled with nitrogen behave. No sooner was it placed in its glass cage than it was overcome as if with stupor. Staggering, beak open, breast heaving, it appeared almost at its last gasp. Seized with convulsions, it fell sidewise, struggled aimlessly, opened its beak again and again as it panted for breath, and then ceased to move. The bird was dead. The other, on the contrary, was still comporting itself in a very lively fashion.

"This experiment," Uncle Paul confessed, "is one that I have no liking for; nor does it please you any better, my dear children. The sight of a creature in pain, suffering as the victim of our curiosity, like this sparrow dying to afford us instruction, is as repugnant to your kindly nature as it is to mine. It is a thing to see once, the pursuit of knowledge having its cruel necessities; but it is not to be repeated. Let us hasten to liberate the survivor. For the sake of its fellow, dead in the cause of chemistry, I forgive it for pilfering my peas."

The two sparrows were taken out of the bottles, the one from that containing air being as lively as a cricket. Emile held it in his hand a moment, bade it good-by, took it to the open window, and let it go, whereupon it flew away like an arrow, with a cry of supreme joy. The other, its poor little claws stiff in death, remained on the table, breast upward. Emile and Jules gave it occasional side glances, puzzled to understand the cause of so sudden a death, and perhaps hoping to see it come back to life. Their uncle perceived what was in their minds.

"Do not hope for the sparrow's revival," said he. "It is dead—poor little thing!—dead for good and all."

"Is this nitrogen, then, such a terrible poison?" asked Jules.

"No, my young friend. Far from being a poison, nitrogen is perfectly harmless. It must be harmless or we could not live in an atmosphere of which it constitutes four fifths. We all breathe it incessantly, and not one of us ever has any reason to complain of it. Nitrogen is quite harmless; that is not what killed the bird."

"Then why did it die?"

"The candle which burns in the air goes out in nitrogen. Is it because of this gas? No, for if that were so, the candle could not burn in an atmosphere abounding in nitrogen. Where this gas is present all by itself the candle goes out, not on account of the nitrogen, but because it lacks the element essential to combustion, and that is oxygen. It is not the presence of the one gas, but the absence of the other, that makes combustion impossible.

"We ourselves perish quickly in water. Why? Can it be that water is a poison? Certainly not; such an idea would never occur to us. We die in the water for want of air, water itself having nothing to do with the death of the person drowned, which is due solely to the lack of breathable air. In like manner we may say the sparrow met its death by drowning in nitrogen. It cannot truly be said that air was wholly lacking, for the bird had a plenty of one of the two gases composing the atmosphere; it was merely deprived of that part which alone is breathable, which alone promotes in the animal, for the purpose of sustaining life, an action comparable in every way to that set up in the candle-flame to cause it to burn.

"It is the lack of oxygen, then, that caused the sparrow's death and the going out of the candle. Where there is no oxygen, neither life nor combustion is possible. No animal can live where a candle cannot burn, for life and combustion are closely akin, as I will show you at the proper time. But first we must study carefully this partner of nitrogen in our atmosphere, this gas called oxygen; and then you will be in a position to see the close resemblance between life and fire."

The boys exchanged glances of surprise at hearing their uncle thus associate these two things.

"I am saying nothing," he continued, "that does not agree with the most careful scientific observations; that has not, indeed, to some extent, become a part of our every-day thinking, so obvious is it to every one. We say of a fire that has gone out, that it is dead. The famous song that *Harlequin* sings under his friend *Pierrot's* window tells us that the candle is dead. In order to die it must first have lived. The dead fire, the dead candle, had, while they burned, I will not say life, for that would be going too far, but at least a state not unlike life in respect to the chemical action concerned. A lighted candle and a live animal consume oxygen in order to continue burning in the one case and living in the other. Both candle-flame and animal die in nitrogen because this oxygen is lacking there. That is the whole secret of the end of the poor sparrow."

"And other animals?" asked Emile. "Would they die in the nitrogen as the sparrow did?"

"All would die there, some sooner, some later, according to their different kinds—absolutely all, since no creature, however small, can live without oxygen, for which nitrogen will not in the least serve as a substitute. If it were not a cruelty as repugnant as it would be useless, we might repeat our experiment with all the inhabitants of our garden,—with its birds, field-mice, moles, insects, snails, and so on,—and we should see them all succumb in nitrogen, some quickly, others after a long enough time to try our patience; for I must tell you that though all animals without exception need oxygen if they are to live, they do not all feel its need with equal urgency. There are some that are overpowered instantly in nitrogen: such was the case with our sparrow; others can live in it for hours or even for days, but are sure to die at last. The rule is universal; the length of the victim's resistance alone varies. First to succumb are the birds, their breathing being very rapid. Then follow the fur-bearing animals,—the cat, the dog, the rabbit, etc.,—in short, the mammals, as naturalists call them. Reptiles have a much greater power of resistance:

a lizard, a snake, or a frog would perhaps not be quite dead even at the end of an hour. Finally, insects, snails, and other small forms of animal life are the last to perish.

"This is something so important that I must not refuse to illustrate it by experiment, despite our pity for suffering. Besides, I have in mind a poor victim that would otherwise perish miserably under the cat's claws. Better that it should meet with a gentle death in nitrogen than endure the cruel suffering the cat's claws would inflict. We shall be doing it a kindness to spare it that torture. It is a mouse caught in the mouse-trap. I saw it this morning on one of the pantry shelves. Emile, go and get it."

Emile came back with the mouse-trap and its captive. From the nitrogen left in the bell-glass, the bottle in which the sparrow had died was refilled. Opening the trap a little, Uncle Paul dropped the mouse into the bottle. Finding itself in this glass prison, the animal first circled around several times, hugging the wall and seeking an outlet, with no further appearance of discomfort than fright. Then it crouched down, began to tremble, and seemed to go to sleep. Finally, a sudden convulsion announced that it was dead. Only a few minutes had passed, but it was clear that the animal had taken longer to die than the bird.

"Give the mouse to the cat," said Uncle Paul, "and that shall end our experiments with animals. Now let us sum up what we have just learned. Nitrogen forms four fifths of our atmosphere. It is a colorless, odorless, invisible gas in which nothing can burn. A lighted candle goes out the moment it is lowered into this gas. Nor can animal life sustain itself in nitrogen: any animal that breathes it unmixed with oxygen dies sooner or later, not on account of the nitrogen itself, which has no harmful properties, but for lack of oxygen, the only part of the atmosphere that will sustain life."

## CHAPTER X

### BURNING PHOSPHORUS

PREPARATIONS had been made for some new experiments, which pupils always like. On the table stood the tin box containing the bottle with the phosphorus; also there was the famous bell-glass resting on a plate, in the middle of which was a saucer full of lime.

"What is Uncle going to show us with all these fixings?" queried the boys.

"The air we breathe," he began, "is still very imperfectly known to you. Of the two elements composing it, only one, nitrogen, has been shown you; the other, oxygen, less abundant but much more important, is hardly known to you except by name. You recall what the experiment with burning phosphorus showed us,—that oxygen forms the fifth part of our atmosphere,—and you also know, rather from my telling you than from the evidence of actual facts, that it is the gas needed when anything is to be burned. Without oxygen the flame goes out, and without it the life of an animal also comes to an end. But what is this gas? What will it do alone, by itself, and not mixed with nitrogen as it is in the atmosphere? That, my little friends, is the important question. I am going to try to answer it for you.

"In five liters of atmospheric air there are four liters of nitrogen and one of oxygen; so that is the source we must go to when we wish to obtain in a pure state either one of these two elements. Now, in the atmosphere the two gases are not chemically combined, but simply mixed, as I shall have occasion to prove to you later. As they are only mixed, a simple separation of the two is all that is needed, though even this is a difficult matter; for how can we separate two substances that cannot be handled or even seen? A little while ago when we mixed powdered sulphur and iron filings, Emile thought it not impossible to separate the two, grain by grain, at a great cost of time and patience. And he was right: the task is not too much for nimble fingers and sharp eyes. With this mixture called air, however, it is a very different thing. The two substances forming the mixture can be neither seen nor felt, and if they could be seen it would still be hardly any easier to separate them, so subtle is their nature. What, then, are we to do?"

"It was easy enough to separate iron and sulphur," said Jules, after a moment's reflection, "by using a magnet, even though the two substances were both of them powdered very fine. Could n't we use some means for sorting the two gases that air is made of?"

"Yes," chimed in Emile, "I 'd like to find something that we could hold in the air and make it attract one of the gases and leave the other behind, just as the magnet attracted the iron filings and left the sulphur."

"Do you know, my lads, that what you say shows more understanding of the matter than I had expected?" rejoined Uncle Paul. "Your answers delight me, anticipating as they do what I was about to propose as the only practicable means to be employed. What Emile says he would like to find is already known to you; you have seen it in operation, and no longer ago than day before yesterday."

"Phosphorus?" queried the boys.

"Yes, phosphorus. When it was burned under the bell-glass, did it not take to itself the oxygen and leave its companion, nitrogen, in the glass?"

"Yes, that 's just what it did."

"Did n't it behave very much like the magnet you thrust into the mixture of iron filings and sulphur so that it drew the iron filings to itself and left the sulphur on the paper?"

"To be sure it did!"

"The magnet attracts iron, but has no effect on sulphur, which is thus left by itself. In the same way burning phosphorus attracts and retains the oxygen in the air, but leaves the nitrogen, for which it has no liking."

"Now we have it, I think," said Jules. "The magnet, covered with iron filings, was drawn out of the mixture, and then we rubbed off the filings on to another piece of paper away from the sulphur. Let 's make the phosphorus take all the oxygen it wants, and then we 'll take it away again."

"A capital suggestion," applauded Uncle Paul;

"but, unfortunately, it won't quite work. The magnet readily gives up its load of filings, but not so with phosphorus and its load of oxygen. I have told you of its voracious appetite. Once having got its fill of oxygen, it is impossible to make it disgorge except by forcible means not at our command in our humble laboratory. What it gets it keeps a good hold of, so that with our modest resources we should never succeed in making it let go."

"Let it keep its oxygen, then!" cried Jules in vexation at seeing his project fail just as he thought it about to succeed. "I 'll try another way. Is n't there something that will work just the opposite of phosphorus,—something that will take the nitrogen from the air and leave the oxygen alone by itself? That would be much simpler."

"No doubt that would be much simpler, but—"

"Is there a but?"

"Alas, yes, and a most serious one! You must know that nitrogen is a most unsociable element, decidedly hostile to the notion of alliances. No element will have anything to do with it, as a rule, and it has no use for a partner of any sort. Chemical combination it abhors, and only when coaxed by the most skilful and delicate devices will it consent to any such union. Let us not, then, for a moment think of withdrawing nitrogen from the air by combining it with another substance; all attempts in that direction would be sure to result in failure.

"Must we, then, give up in despair? Not at all. The first method is excellent if we only use it with discretion. Phosphorus, it is true, keeps a most obstinate hold on the gas it has united with in the act of burning, and it is useless to expect it to let go of the oxygen it has taken from the air. But, fortunately, not all simple substances are like it. We shall find some more accommodating, willing to surrender their plunder without too much coaxing. For to-day we will content ourselves with learning how this gas is accumulated and, as we might say, stored up in a burnt substance; and for the purposes of this demonstration phosphorus will serve.

"You have not forgotten how smoke formed in the bell-glass when phosphorus was burned there, in our experiment of day before yesterday. That thick cloud of milk-white appearance made too deep an impression on you to be soon forgotten. And you remember how, little by little, it dispersed, being taken up by the water in the bowl. If I had not called your attention to this point, perhaps that disappearance would have seemed to you a real instance of annihilation, and you would have retained the notion, so generally held, that fire reduces to nothing the material that it burns. I instructed you to the contrary, but that is not enough; I wish to add to my mere assertion the more convincing testimony of fact. Accordingly, I propose to show you that fire does not annihilate, but only transforms; that it changes the appearance and the properties of matter without affecting its existence. Phosphorus will furnish us a fine example of this, and at the same time give us some knowledge of the chief topic of our to-day's lesson. The experiment I am proposing will show us, on the one hand, the indestructibility of matter by fire, and, on the other, the storing up of oxygen by combustion.

"The white fumes given out by burning phosphorus are very easily dissolved in water, which accounts for their prompt disappearance in our recent experiment. To preserve them, to let them take on the state natural to them when cold, and then to examine them at leisure, it is absolutely necessary to do the burning where there is no water. Nor is even this precaution sufficient, so great is the liking for water of this compound formed by burning phosphorus. The atmosphere is always moist, whence come rain and dew. However dry it may seem to us, it is sure to contain more or less of the invisible vapor of water, which the burned phosphorus would greedily pounce upon, dissolving itself therein just as sugar dissolves in water. Consequently, we must have perfectly dry air in the bell-glass where the burning is to take place.

"This dry air I obtain by means of quicklime,—that is to say, lime before it is slaked by the mason, or, in other words, lime just as it comes from the lime-kiln where it has been prepared. You don't need to be told what happens to a piece of lime left in the air for some time."

"I know what you mean," said Jules. "The piece of lime gradually cracks open, and then crumbles to dust, just as it does when you sprinkle it with water; only then it crumbles a good deal faster."

"That is it. Sprinkled with water, a piece of lime cracks, splits, crumbles to dust. Exposed to the air for some time, it acts in the same way, but more slowly. Why? Because it absorbs the moisture in the surrounding air, until, little by little, this moisture has had the same effect on it as a fine spray of water would have had. Thus it is that lime has the habit of attracting moisture, however little there may be within its reach. It wrenches it by force, we might say, from the surrounding atmosphere, and takes every bit of it. Here we have, then, a very easy and convenient way to obtain perfectly dry air.

"A few hours ago I took care to place a saucer of quicklime in the middle of a large plate, and covered it with the bell-glass, the latter resting on the plate and being, of course, full of air. With this precaution the fumes of the burning phosphorus cannot escape, and not only the imprisoned air, but also the surface of the plate under the glass and the inside of the glass itself, will be rendered very dry. Now for the burning of our phosphorus."

Uncle Paul cut off a piece of this substance under water and dried it carefully with blotting-paper. Then, placing it on a bit of broken crockery, he withdrew the saucer full of lime and substituted the phosphorus, which he set fire to, whereupon the bell-glass with its contents of dry air was immediately replaced on the plate. The burning did not at first differ in any way from that already seen by the boys; there was the same bright light, with the same eddies of dense white smoke. But at this point there occurred something new: on the inside of the cold glass the smoke condensed and became a beautiful, white, flaky substance that detached itself and fell here and there with all the appearance of falling snow. Soon the plate was covered with a layer of this strange snow-like substance that had come from the very heart of the flames.

"Well, Emile," said his uncle, "what do you think of this snow?"

"I think it is very wonderful. Who would ever have expected fire to make a snow-storm! But I know well enough it is n't real snow, though one might be fooled by its looks. Those flakes, all so white and beautiful, must come from the burning phosphorus, for they could n't come from anything else."

"Yes, that is quite clear. The substance we see forming here has nothing of snow about it except its appearance. In reality it is quite another thing, as we shall soon see. But first let us make it snow a little more. The fire is dying down; we will feed it."

Uncle Paul raised the bell-glass a little, and the burning, which had begun to languish, started up again with all its former vigor.

"Air was beginning to fail," said he; "the phosphorus had nearly exhausted its supply of gas, and was about to die down, but, by raising the bell-glass a little, I have let in more air, and the fire revives. Let us give it a little more air still, so that we may be sure to have enough of this strange snow."

When, after three or four renewals of air, the layer of snow on the plate was deemed thick enough, Uncle Paul took a pair of pincers and drew out the piece of crockery with the phosphorus on it and carried it into the garden, that the still unconsumed material might there burn itself out with no inconvenience to lungs or sense of smell.

"Now, my young friends," he resumed, "I invite you to examine what there is in the plate. It is, as you see, a white, flaky substance looking much like snow. That is what phosphorus turns to when burned. Fire has not destroyed it, but has changed it into something else, and the change is so complete that if you did not know where this false snow came from you could never guess its nature. I repeat, fire does not destroy anything; what it devours, what it consumes, is not reduced to nothing, but changed into something else, which sometimes vanishes from before our eyes as an invisible gas, and at other times arrests even the least heedful attention as a much grosser substance. What you see here in the plate—this stuff that we can feel, smell, and taste—is phosphorus consumed by fire, phosphorus still in existence though it has been burnt up. Thus is illustrated before your eyes the first point I had in view in this experiment,—namely, that nothing is ever annihilated even by the action of fire.

"Suppose we had here a fine pair of scales for weighing, exactly balanced as are scales used by chemists, and capable of telling us with precision the weight of a fly's wing. The weighing of even so minute a bit of matter is very often called for in the delicate operations of chemistry. With a pair of scales of such sensitiveness, we could have ascertained the weight of the piece of phosphorus in milligrams. Nothing then would have stood in the way of our burning the whole of that piece under the bell-glass by renewing the air as often as was necessary; and at the end we could have taken a feather and swept up the snowy deposit even to the last flake, after which we could have weighed it on our scales. Let us suppose these two weighings to have been carried out, one before the phosphorus was burned and the other after. Now, which will weigh the more, the unburnt substance or the burnt?

"Misled by that false notion of fire as a destroyer, the novice would answer at once that the burnt substance would weigh less than the other, arguing that if fire does not entirely destroy, it at least destroys in part. But you, my boys, forewarned against this error in our earlier talks, and having had your eyes opened by a number of experiments, will not, I believe, make this foolish answer."

"I should think not," was the confident reply of Jules. "I should say the burnt phosphorus would be heavier than the unburnt."

"Your reason, my lad. Let us make no assertions without proof to support them."

"The reason is plain enough," said Jules. "You told us and proved to us that when anything is burned it combines with something in the air called oxygen. Although an invisible gas, this oxygen is matter, and consequently has some weight, if only a very little. So that burnt phosphorus, having had oxygen added to it, ought to weigh more than phosphorus alone."

"Golden Mouth could not have said it better," applauded Uncle Paul. "Yes, my young friend, the burnt phosphorus must weigh as much as it did before combustion, plus the weight of the gas combining with it in burning. A delicate pair of scales would testify to this in the most convincing fashion: it would show us that this heap of what looks like snow-flakes weighs more than the phosphorus that went to the making of it. How account for the added weight except by ascribing it to the air that played its part in the combustion? So, then, in the substance on this plate, in the burnt phosphorus, there is a small amount of oxygen taken from the atmosphere and securely retained. This oxygen has ceased to be an invisible gas occupying a great space, and has become part of a solid substance that can be seen and handled, and that occupies a comparatively small space. It is stored up as in a reservoir, where chemical combination has collected and compressed it into the least possible bulk.

"A similar chemical action attends the burning of any substance whatever. By being consumed it becomes a storehouse of oxygen. Taken in its sum total, and with no omission of any part, the resulting matter after combustion is heavier than the substance before combustion; and this excess of weight is due to the gas that took part in the combustion. Most of these burnt substances, veritable storehouses of oxygen, hold on to the latter with a tenacity that can be overcome if necessary, but that nevertheless offers very great resistance, while the lesser number surrender this oxygen easily. After a brief survey of this latter class we will select the substance best suited to our purpose of obtaining oxygen in its pure form. But we will first finish our examination of burnt phosphorus while we have a specimen before our eyes.

"Although derived chiefly from phosphorus, which is highly inflammable, the snow-like powder in the plate is absolutely incombustible, the hottest fire having no effect on it, for the reason that what is once burned cannot be burned again. This phosphorus, being already combined with as much oxygen as its nature permits, cannot take on any more; that is, its combustibility is at an end. Experiment will prove this better than mere words."

On a pan of glowing coals from the kitchen fire was sprinkled a little of the white powder, and the coals were then blown to a more intense heat; but the powder showed no signs of taking fire, its inflammability having been quite exhausted.

"If you had not already," Uncle Paul resumed, "some knowledge of the difference between compound substances and the simple substances composing them, this experiment would open your eyes, for the substance that at first burned so freely now refuses to burn at all. Let us proceed. You can convince yourselves that the white powder in the plate has no smell whatever, whereas the phosphorus had a very strong smell of garlic. But I would not have you touch this powder, its properties being such as to render handling harmful; still less do I wish you to taste it, as it would make you cry out with pain."

"Is it so terrible as all that?" asked Emile.

"So terrible that a drop of molten lead on your tongue would be less painful."

"But that snowy stuff looks harmless enough."

"Don't trust to looks, my little friend. Innocent looks may disguise a very dangerous substance. Forewarned is forearmed. From the kitchen of the chemist there very rarely comes anything pleasant to the taste. However, it is well for you to have some idea of how phosphorus tastes, and to make it less disagreeable to the tongue I will dissolve it in water."

So saying, Uncle Paul took up the feather again and swept the contents of the plate into a glass of water. As each particle fell into the liquid it gave a hissing sound, like that produced when a blacksmith plunges red-hot iron into water.

"It must be awfully hot, must n't it," asked Emile, "to hiss like that in the water?"

"Heat is not the cause of the hissing. The powder is no hotter than anything else here,—no hotter than the plate it lies on. I have already told you that burnt phosphorus has an extraordinary liking for water; and you know what extreme precautions I had to take, with the aid of quicklime, to keep it from the moisture in the atmosphere. Now I let this powder drink as much water as it pleases, and it dissolves immediately and even with violence, that sharp noise testifying to a satisfied thirst.

"Behold, now, the snowy powder all dissolved in the water. The liquid has not changed its appearance; it is still water as far as looks are concerned; but dip the tip of your finger in and taste it. You can do so without the least fear."

As the children hestitated, remembering the allusion to the drop of molten lead, their uncle dipped the tip of his little finger into the liquid and touched it to his tongue; whereupon, emboldened by his example, Emile and Jules did likewise.

"Oh, how sour it is!" they cried, surprised at the disagreeable taste and making a wry face. "It is sourer than any vinegar they use for making salad. What would it have been like if Uncle had n't weakened it with a lot of water?"

"Your tongues would have suffered terrible tortures, my little friends. The part touched would have been eaten away at once by this violent chemical, and you would have heard a hissing as of red-hot iron in contact with your saliva."

"Then this strong vinegar is n't real vinegar?"

"It is not vinegar at all, though it tastes much like it. Now let us go on. Phosphorus has still another property that we must test. Here are some violets just gathered from the garden. I dip one into the sour-tasting liquid, and it immediately loses its blue color and turns red. All flowers of the same color as the violet—the iris and the bluebell, for example—would likewise turn red in the sour liquid. You shall at your leisure repeat this fine experiment with all the flowers you can find in the garden, and you will always see the blue flowers turn red. Burnt phosphorus, then, always has this sour taste and this quality of turning blue flowers red.

"I will add that most of the other metalloids, such as sulphur, carbon, nitrogen, and many more, when they combine with oxygen,—or when they burn, as we say,—produce compounds of like sourness and of like ability to change the color of blue flowers to red. All these compounds are called acids, from their sour or acid taste, and they are distinguished one from another by the addition of a second term indicating their origin. Thus the snowy powder resulting from the burning of phosphorus bears the name of phosphoric acid, and in future that is what we will call burnt phosphorus.''

# CHAPTER XI

## BURNING METALS

ALL the blue flowers in the garden had been tested with phosphoric acid. After the violet had come the iris, after the iris the periwinkle, then the veronica, and others besides; and all had lost their natural color and turned red. Yellow flowers, on the other hand, such as the marigold, and white ones like the Easter daisy, and red ones such as the poppy, did not change color at all when dipped into the acid liquid. For some time, now, the contents of the glass had been used in these experiments when Uncle Paul invited his pupils to take part in some new investigations; and the summons was responded to with an outburst of joy. This time the equipment consisted of a small portable stove full of charcoal and set to burn in the fireplace, while on the table were some bits of zinc, doubtless the remnants of a worn-out watering-can. There was also an old iron spoon, in so sorry a condition that no beggar would have stopped to pick it up in the street. A bottle hardly longer than your finger contained something grayish, of metallic appearance, shaped like a narrow ribbon and wound about in the form of a skein. The boys puzzled in vain to guess what this could be. Their uncle told them in due time, but before doing so he resumed his talk as follows:

"How to obtain pure oxygen freed from the nitrogen with which it is mixed in the atmosphere was the difficult problem engaging our attention at our last lesson, and it will continue to do so to-day. We know that the sour-tasting compounds, the acids, made by burning various metalloids, notably phosphorus, contain an abundance of oxygen taken from the air and stored up. That is the first stage of the journey toward our destination. Another lies ahead, and when we have accomplished it we can, with better understanding, avail ourselves of the means offered us by chemical science. Perhaps you will say I am straying from the road that leads straight to the element, oxygen, which you are so eager to learn about, and you will reproach me for dividing the journey into more stages than necessary."

"Oh, never fear," Jules hastened to reply; "you may make the journey as many stages long as you like. If they are all as interesting as the last one, when you showed us snowflakes made by fire, we sha'n't be the ones to complain. Oxygen will have its turn all in good time."

"To-day's stage in our journey, my young friends, will be no less interesting than the one that preceded it. I think it will give you an even greater surprise; and in the end it will show us how to obtain pure oxygen, which is what we started out for. Let us talk a little more about the burning of various substances.

"A piece of phosphorus on fire is certainly a fine sight. The vigor with which it burns, the dazzling brightness of its flame, the snowy flakes of phosphoric acid resulting from the combustion—all these cannot fail to arouse interest. But, accustomed as you already were by the use of matches to seeing phosphorus burn, this spectacle had in your eyes none of the fascination of the new and unexpected. To witness the burning of a substance well known to be highly inflammable is nothing very thrilling. But to-day you will be rather surprised to see things burn that you have always thought to be fireproof. We are going to set metals on fire."

"Metals!" cried Emile with a start of astonishment.

"I said that would rather surprise you. Yes, my boy, metals, real metals."

"But metals don't burn."

"Who told you so?"

"Nobody told me, but I know what I see every day. Tongs, fire-shovel, andirons are made of iron, which is a metal; and, even in the hottest kind of a fire, I have never seen any of these things burn. The stove is metal, and in winter when it gets red-hot with a roaring fire inside I have never seen a particle of it burn. Why, the whole stove would have burnt up long ago if metals could take fire as you say they do!"

"Then Emile does not believe me when I say that metals can burn?"

"What can I say, Uncle Paul? You put my faith

in you to a hard test. You might as well tell me that water can burn, too.''

"And why not, as I am going to show you some day that it can?''

"Water burn?''

"Yes, my child; I propose to show you some day that water contains the very best kind of material for burning.''

Astounded at this promise, made with undoubted sincerity, Emile said no more, but waited to see metals burn, before believing what seemed to him incredible. His uncle continued:

"If the tongs, shovel, andirons, stove, and other things made of iron do not burn in our fireplaces and kitchens, it is because the heat is not intense enough. Make more heat, and metal will not fail to take fire. You have often seen this burning of iron, but without suspecting what it really was. Let us recall what we see from time to time at the blacksmith's shop when we pass his door. The smith has just taken from the forge a bar of iron, glowing hot. As soon as it comes to the air this bar of dazzling brightness throws out showers of brilliant sparks in every direction, so that one might mistake it for a piece of fireworks. The dark shop is all ablaze with blinding flashes. What are these sparks that fly off in showers? Little scales of iron that become detached from the bar and burn as they shoot through the air. Does Emile begin to believe me now?''

"Yes, I am beginning to. I see more and more every day that anything may be expected in chemistry.''

"I will tell you further that when the makers of fireworks wish their pinwheels, Roman candles, rockets, and squibs to throw out splendid sparks like sprays of water from a fountain, they mix with gunpowder a quantity of filings of various metals according to the colors they wish to obtain. Copper gives green sparks, iron white. Each particle of these metal filings turns into a spark as soon as it touches fire. I am planning to show you, one of these days, a conflagration of iron that will make the blacksmith's shop look dull and tame. So I will say but little more at present about this metal, merely adding a third example to the two I have just cited.

"You both know how to get bright sparks by striking flint and steel together, or flint and the back of a knife-blade. Those sparks are particles of metal, of steel, which is a variety of iron, that are struck off, becoming heated by the shock and burning as they fly through the air. In exactly the same way sparks fly off from the stone of the scissors-grinder when he sharpens a piece of cutlery on it, and from an iron horseshoe when it strikes a pebble. The violent friction of the grindstone and the shock of the horseshoe's hitting the hard pebble break off fine scales of iron, and these, raised to intense heat by the friction, catch fire as they dart off through the air. You see, you don't have to go very far to find plain proofs that iron really will burn, which at first seemed so impossible to you. The scissors-grinder sharpening a knife and the mule stumbling on a stone teach us what Emile found so hard to believe;

they perform for the wide-awake observer one of chemistry's fine experiments.

"I pass now to another metal, zinc. Here are some pieces, the remains of an old watering-can that has been thrown away. This metal is grayish on the surface, but if I scratch it a little with a file or the point of a knife, we get a glimpse of metallic luster like that of tin or silver. Now what we wish to do is to make this zinc burn,—quite an easy operation and readily performed with the help of a few live coals. It is the same with metals as with the things we commonly think of as inflammable, such as sulphur and phosphorus and charcoal, for example: some catch fire easily, others with difficulty. At the mere touch of a lighted match phosphorus is instantly ablaze, while sulphur is much slower and charcoal more obstinate still. In the same way, while iron requires the heat of a forge to make it burn, zinc needs only a few live coals. There are some metals that burn still more easily, as we shall soon see.

"Let us proceed, now, to make zinc burn. I place a few pieces of it in this old and discarded iron spoon, and thrust the spoon thus filled into the midst of the glowing coals in this little stove. If you have any doubts, our experiment will settle them for you."

Everything being arranged as Uncle Paul had indicated, they waited a little while. The zinc melted almost as readily as lead, and when the spoon was red-hot the coals were pushed aside a little to show the melted zinc without cooling it off. Then with a stout piece of iron wire Uncle Paul began to stir the molten metal so as to bring it into better contact with the air. A flame burst forth, of a superb bluish white, dazzling to the eyes, though of no great volume. It flickered on the surface of the liquid metal, springing up brighter or dying down, according to the briskness with which the matter was stirred. The boys marveled at the brilliance of the burning zinc, and their wonder was heightened as they saw a kind of snow rise from the flame, float lightly in the air, and spread all about the room. One might have taken it for the finest of incomparably white down, or those delicate shreds of cobweb that one sees fluttering over the fields on a fine autumn morning. At the same time, on the surface of the metal in the spoon, there collected a kind of cotton-wool of unparalleled fineness. The current of hot air from the stove moved it lightly, detaching the flakes they saw rising and spreading through the room.

"This white fluff," resumed Uncle Paul, "this down or cotton, is burnt zinc, zinc combined with oxygen from the atmosphere, and it bears the same relation to this metal that the snowy flakes bear to phosphorus. We will ascertain its chief properties as soon as there is enough in the spoon."

Jules took his uncle's place in stirring the molten metal, while Emile began to blow the flakes before him, taking care not to blow hard enough to drive them too fast, even the largest of them floating so lightly in the air, despite their size, that it seemed as if they would never settle down. Soon there was no more of the shiny liquid left in the spoon, all the zinc having turned into this white material. When the residue of this in the spoon had cooled off and been emptied out, Uncle Paul continued:

"Burnt zinc is a white substance such as you now see before you. It is also quite tasteless. Put some on your tongue and it will convey no sense of taste."

"That 's so," affirmed Emile, after trying it cautiously, warned by his previous experience with burnt phosphorus. "It has n't any more taste than a pinch of sand or sawdust."

"I don't taste anything, either," Jules chimed in; "and yet the burnt phosphorus, the phosphoric acid, was so sour you could n't bear it. Now, here 's something else that we 've burnt, and it has n't any taste at all."

"Let us seek the reason for this lack of taste," suggested his uncle. "I drop a good pinch of the white substance into this glass of water and stir it well with a stick. It does not dissolve, does not melt in the water; it is, as we say, insoluble. You remember how readily, on the other hand, the burnt phosphorus or phosphoric acid dissolved in water."

"We are not likely," replied Emile, "to forget very soon that terrible stuff that melts in water with a hissing sound like red-hot iron, and that has to have dried air if you want to keep it."

"Let us put together these various facts: burnt phosphorus dissolves in water and has a taste; burnt zinc does not dissolve in water and has no taste. In the same way, salt and sugar readily dissolve in water, and both have a taste, the first a salt taste, the second a sweet taste. Marble and brick do not dissolve in water, nor has either of them the slightest taste. Do you begin to see what these facts point to?"

"It seems to me from all that," replied Jules, "that to have any taste a thing must be able to melt in water."

"I have no fault to find with that answer. Yes, my boy, what a substance must have in order to possess any taste at all, whether strong or weak, sweet or sour, salt or bitter, or whatever flavor, is solubility in water. Anything that cannot be dissolved in water is by that very fact tasteless, and for this reason: to act on the sense of taste, to make an impression on tongue or palate, a substance must necessarily, unless it be a liquid, dissolve in the saliva, being thus divided into extremely small particles and brought into contact with the organs designed to perceive it. Now, saliva is almost entirely composed of water. If, then, a substance is insoluble in water, it is also insoluble in saliva, and therefore has no taste. In future, when you see a substance that will not dissolve in water, don't try to see how it tastes, for it has no taste and cannot have any. But if it yields to the action of water, it has a taste,—sometimes very insipid, it is true, and hardly perceptible, as gum Arabic, for example.

"To return to our two burnt substances: the white material left by the burning zinc has no taste, because it is not soluble in water, whereas the white material left by the burning phosphorus is easily soluble and consequently has a very decided taste."

"Yes, very decided," assented Emile, "for it eats

away the part of the tongue it touches. But tell me, Uncle, if burnt zinc could melt in water, so as to have a taste, what sort of a taste would it be? Would it be as strong as the taste of burnt phosphorus?"

"As to that, my little friend, neither I nor any one else can give you a sure answer, no one having experimented with such an impossibility. All we can say is that probably the taste would be detestable, as is the taste of ninety-nine chemicals out of a hundred.

"When we exhibit fireworks we keep for the last the most beautiful piece of all, the star number on our program. This I am doing to-day, reserving to the end the prize piece in our collection, the most splendid example of metal-burning. The material for this display is there in that little bottle."

"That thing that looks like a skein of narrow gray ribbon?" asked Emile.

"Yes, that 's it."

"It does n't look as if it could do anything."

"It can do much more than its appearance promises. Let us examine it closely."

So saying, Uncle Paul took the skein out of the bottle. It was a dull gray ribon, narrow, very thin, and as flexible as tinfoil. When scratched with a knife it showed the brightness peculiar to metals. By this brightness and the whitish color the children thought they recognized the metal.

"It is either lead or tin," declared Emile.

"I should rather say zinc or iron," said Jules.

"It is not any of those metals," their uncle told

them. "This is a metal you have never seen before, nor have you ever even heard of it."

"And what is it called, please?" was Emile's eager inquiry.

"It is called magnesium."

"Ma-ma—" stammered the boy. "Please say it again."

"Magnesium."

"Oh, what a funny name!"

"No funnier than bismuth, barium, or titanium."

"Are those the names of metals?"

"Yes, my child, they are the names of metals. If they sound strange to you, it is because you now hear them for the first time. One gets used to bismuth and titanium just as much as to copper and lead. As I have already told you, there are about fifty metals. Most of them remain unknown to us, not being in common use; their names, being seldom heard in ordinary conversation, strike the ear rather strangely at first. After you have become acquainted with magnesium, you will find its name easy to remember and it will cease to have the unwonted sound it has at present.

"A few live coals are enough to make zinc burn; a candle-flame will set fire to magnesium, and the metal will go on burning of its own accord when once it is started. It lights almost as easily as a paper spill."

"And where do they find this queer metal?" Emile inquired. "I should be willing to spend some of the pennies I have saved up if I could buy a piece."

"Magnesium is not a metal in common use. It is

unknown to the blacksmith, the tinsmith, and the coppersmith. It is a substance used chiefly in scientific research and in chemical experiments of an entertaining sort. It can be procured at city drug shops and toy-shops, where it is sold as one of those curiosities that instruct and amuse at the same time; and it is from the city that I brought it home for your benefit."

A candle having been lighted, the shutters were closed so as not to let the daylight lessen the brilliant effect of the burning metal. Then Uncle Paul cut off a short strip of the magnesium ribbon, and, taking hold of one end with a pair of pincers, touched the other to the candle-flame. A sheet of paper was spread on the table to receive what might fall from the burning metal. Catching fire very quickly, the ribbon was withdrawn from the candle and held in an upright position over the sheet of paper, after which it needed no further assistance; the magnesium burned alone. It was as if a sudden burst of dazzling sunlight had illuminated the darkened room. A superb white light, intense enough to penetrate every corner and render all objects clearly visible, came in waves from the splendid torch. No sputtering, no noise whatever, no flying sparks. It was the calm, sustained illumination of daylight. Dumb with astonishment at this brilliant display, the boys gazed in fascination. The burning continued, the flame approaching ever nearer to the pincers, while the burnt part fell off in pieces that looked like chalk. In a few seconds it was finished; the radiant flame died down for want of fuel.

"Oh, how beautiful, how splendid that was!" cried the children, rubbing their eyes, dazzled by the glare.

Their uncle opened the shutters to let in the daylight.

"Why, I can't see;" said Emile, still rubbing his eyes. "I 'm nearly blind from looking at the magnesium flame.

"And, I," added Jules, "am almost as dazzled as if I had been staring at the sun."

"That will pass off in a few minutes," their uncle assured them. "Wait until your eyes recover from the fatigue caused by the too-bright light of the magnesium flame."

The dazzling effect having worn off, as it did in a short time, Emile spoke of something that had struck him while the magnesium was burning.

"I was looking," said he, "at the flame of the candle, which you had left burning after setting fire to the magnesium, and all I saw was something of a reddish color, smoky and dull. How pale the candle looked, and it had been bright enough before! I could hardly make out any flame at all. Is it possible, I asked myself, that that can give light?"

"If you set a lighted candle in the full glare of the sunshine, can you see the flame?" asked Uncle Paul.

"No; it looks as pale and dim as it did in the magnesium light."

"That, my little friend, comes from the inability of the eye, when exposed to a bright light, to see, except imperfectly, a dim one. In full sunlight one cannot tell whether live coals are really alive or not. The flame that gave light in the dark ceases to do so when overpowered by a greater brilliance. Our dazzled eyes and the apparently dulled flame of the candle prove to us that the magnesium light is one of the brightest possible; it bears comparison only with the sun.

"I hope I have now convinced you, including even the skeptical Emile, that metals are not hard to burn. The flying sparks of burning iron in the blacksmith's shop, the firing of zinc in our old spoon, and finally the dazzling magnesium flame, have furnished proof upon proof. Furthermore, the last experiment shows us that among the metals there are some that would give us splendid light if it were not for their price and scarcity. Instead of lamp oil or candle grease we could use a magnesium ribbon for lighting. Who knows what the future may have in store for us in this field? The history of chemistry is full of wonderful discoveries, and we already owe so many and important improvements to this science that we may well look to it for still greater things.

"But without dwelling longer on the brilliant splendors of magnesium let us see what has become of the metal after burning. The substance that has fallen on to the paper over which the ribbon burned is a white material which, when touched, crumbles into a soft powder like flour or, better, like ground chalk of a very fine quality. It does not dissolve in water, and consequently has no taste. In addition to the metal itself, it contains, as does everything else after burning, oxygen taken from the air in the process of burning. So here is another storehouse of oxygen from which this gas could be obtained by using the proper means, but not without some difficulty.

"Iron burns. Hammered on the anvil when red-hot, it gives out sparks which are tiny scales of this metal on fire. Let us get some of these particles of burnt iron from the smith, and we shall find them to be of a black material, rather hard, but still yielding to the pressure of our fingers. This black material, this burnt iron, is called oxid of iron.

"Zinc burns, being changed by the process into a white substance of which part is wafted upward from the flame and floats in the air like shreds of cotton or down. This white substance, this burnt zinc, is called oxid of zinc.

"Magnesium burns, and becomes thereby a white substance looking much like very fine chalk and extremely soft to the touch. This chalk-like substance, this burnt magnesium, is called oxid of magnesium.

"As a rule, metals are combustible, though there are exceptions; and in burning they combine with oxygen from the air or elsewhere, being thus changed into compounds which have none of the luster of metals and which take the name of oxid. An oxid is a burnt metal, just as an acid is a burnt metalloid, and both contain oxygen."

## CHAPTER XII

### SALTS

THE white substance left by the burning magnesium had been wrapped in paper, and Uncle Paul took it up again as his opening topic in the next lesson.

"Judging from looks only," said he, "we compared this powder to flour and also to very finely ground chalk. If we considered its properties, we ought rather to compare it to lime. This latter, at first a rough and shapeless stone, swells on being soaked in water, and then cracks to pieces and turns into a white powder closely resembling burnt magnesium. Nor is this resemblance misleading, for chalk is also a burnt metal."

"A burnt metal!" repeated Emile, in surprise. "I never knew chalk was made by burning a metal."

"Naturally not," replied his uncle, "for that is not the way it is obtained. If, in order to get lime, we had to burn the metal that enters into its formation, the mason would never venture to use a trowelful of mortar, because of its excessive price."

"I know how lime is made," put in Jules. "Out in the country they pile up stones and charcoal in a lime-kiln, and the fire bakes the stones and turns them to lime all ready for use."

"That is it. The stone they use is called limestone, and it contains lime and something else; this latter the fire drives out, so that, after the burning is finished, the lime is left by itself, ready for the mason and very different from the original stone. It is really a burnt metal, though the lime-burner knows nothing about the burning of metals, and does not even suspect the presence of a metal in the product of his kiln. The honest lime-burner would be greatly surprised to be told that his lime contains a metal; he would laugh in your face and say you were joking. But nothing could be further from joking; lime comes from the burning of a metal or, in other words, from the combining of a metal with oxygen. Its particles may be likened to the tiny scales that fall from red-hot iron as it comes from the forge, and to the flakes that fly off from zinc burning in the bed of glowing charcoal, and also to that floury powder left on the sheet of paper by the brilliant magnesium flame. In a word, lime is an oxid. "It is true that in producing this oxid the hand of man played no part, the combustion having taken place of itself, perhaps at the very beginning of things; and, since the creation of the world, the metallic part of lime has never been found alone anywhere in nature. It abounds nearly everywhere, but so greatly changed, so completely disguised in the combinations of which it forms a part, that it has required much shrewdness on the part of science to guess its existence, and still greater skill to restore some particles of it to their primitive state, that of a real metal. Here is a pinch of burnt magnesium, and here one of powdered lime; look at them well and tell me, if you can, in what the two differ."

"We can't see any difference," the boys agreed, after a careful examination. "Both are white, and both look like flour."

"I see no difference, either," their uncle rejoined. "We are all three of one mind as to this close resemblance, though we know the two to be different substances. Let us, then, say, with those who are wiser than we, that this powder, lime, is the oxid of a metal, just as this other powder is the oxid of another metal, magnesium."

"What is this lime metal called?" asked Jules.

"It is called calcium."

"Why, that is n't so very different from what the farmers round here call lime; they call it *cals*."

"I will tell you, my boy, that the common speech of our part of the country, like that of other southern provinces, contains the relics of a tongue spoken long ago, venerable relics that, instead of being ridiculed as they are by the foolish, should be preserved with veneration. Our popular terms reproduce, though corrupted by centuries of use, the superb Latin tongue that came from Rome with civilization. In that language lime was called *calx*, and this word has been preserved almost unchanged in the common term you just named. So, you see, the rustic term needs no apology, its parentage being one of the noblest. He who speaks the native tongue of the South speaks something very like

Latin. Men of learning, who delight, not without reason, in going back to ancient usages, have made *calx* into *calcium,* a sonorous and pleasing term, to designate the metal that enters into the formation of lime; but they might just as well have started with our popular word, *cals.*

"Well, then, the metal element in lime is calcium, the name coming from the Latin word *calx,* which means lime."

"Will you show us some of this metal?" asked Jules, eagerly.

"Alas, no, my lad! Our humble laboratory cannot afford so costly a curiosity. Not that calcium is rare, for it is to be found almost everywhere, the very rocks in many regions abounding in it, and whole mountain chains being largely made of it; but the difficulty is to extract it from the compounds that contain it, to bring it back to its original state of a simple metal. If we were to ransack all the chemical laboratories in France, we should perhaps find not more than a handful or two, so difficult and expensive is its preparation. That is why your Uncle Paul's collection of chemicals is now and always will be destitute of such treasures. But, at any rate, I can tell you about it. Imagine something white and shiny, almost like silver in appearance, and as soft as wax, so that it can be kneaded and molded with the fingers. That is calcium."

Here Emile interrupted his uncle. "What!" he cried, "calcium a metal that can be kneaded with the fingers like a piece of soft wax or a lump of clay?"

"Yes, my child, this curious metal is soft enough to yield to the pressure of the fingers and be molded at will."

"I wish I had some to make into little men that would look like silver statues."

"It would be a very expensive amusement, and I have told you why. Also, it would be hard on your hands, for this terrible stuff catches fire more easily than anything you have ever seen. If, while you were molding it, your little statue were all at once to burst into flames, what would become of the sculptor?"

"That would n't be so much fun, would it?"

"Beware, then, of burns! Calcium catches fire by merely touching water. Burning coal, sulphur, and phosphorus are extinguished by water; calcium, on the contrary, is set on fire by it. Let it get a little damp and, behold, it begins to burn. Do not look incredulous when I tell you this, for it is nothing but the exact truth. In a lesson soon to follow I will show you that water is not, as you think it to be, always effectual in putting out a fire; and who knows—perhaps—well, we will see whether my purse will allow it."

"What is your purse going to allow?"

"The purchase, it may be, of a metal that, like calcium, has the peculiarity of catching fire in water."

"There are others, then?"

"Yes, three or four."

"And will you really show us one of them?"

"I make no promise. I will see, I will do my

best, if you continue to take as much pleasure in these studies as you have taken so far."

"We should be hard to please if we did n't like to see magnesium burn, phosphorus and zinc turn to snow, and other metals catch fire in water."

"To return to Emile's wish, calcium's readiness to take fire on touching water being now known to you, you see how dangerous it would be to mold it in your fingers, damp as these always are with their natural moisture. If we had any calcium, it is not a thing to hold in the hands and knead with the fingers; it is a dangerous substance to be left quietly reposing in its bottle.

"But now let us pass on from calcium to its oxid, lime. We find that this has a taste of its own, which the oxid of iron, zinc, and magnesium have not; and this taste is strong, disagreeable, and of a burning nature. The taste of phosphorus, after combustion, is sour, acid, whereas that of burnt calcium is caustic, or fiery. Furthermore, it is not the sense of taste alone that lime affects unpleasantly; it exerts its corrosive energy on every part of the body. Handled carelessly for any length of time, it would gnaw the hands. Like the metal it comes from, it is a harsh substance, and prolonged contact with it is to be avoided.

"Not lacking in flavor, lime ought to be soluble in water; and in fact it is, though not to any great extent, but merely enough to give the water an unbearable taste. If we dilute in water a little lime previously reduced to a paste, the liquid will turn white like milk; then, after it has stood a while, the undissolved lime will settle to the bottom and the water will resume its former clearness; but in this clear water, this water showing no trace of any foreign substance, there is dissolved lime, just as there is dissolved sugar in sweetened water. This we know from the burning, lime-like taste of the liquid."

Illustrating by experiment while he talked, Uncle Paul made his hearers taste of the water in which he had dissolved a little lime. A drop of the liquid on the tip of the finger and touched to the tongue was enough to convince them of its disagreeable taste. Emile made a grimace of disgust, spitting several times and rather overdoing his show of repugnance, for the flavor is not so offensive as his actions indicated. Uncle Paul then continued:

"Here are some violets that I picked just now in the garden. I showed you how these flowers and others of the same color lose their blue and turn red when acted upon by a burnt metalloid, an acid, notably phosphoric acid, which you have experimented with to the nearly complete devastation of the garden, so far as its blue flowers are concerned. Now, what will happen to violets when acted upon by a burnt metal, an oxid? That is what this lime here is going to show us."

Uncle Paul selected a violet and pressed it lightly between his fingers with a little moistened lime, whereupon the flower instantly turned a brilliant green.

"Chemistry seems to be a dye factory," observed Emile, astonished at this new transformation.

"You took a little acid and made the blue of the violet turn red, and now you take lime and make the same blue turn green. When I know enough about chemistry I 'm going to make no end of colors for painting my pictures."

"You will be able to make as many as you please, for chemistry teaches, among other things, how from a colorless substance a bright color may be obtained by certain combinations of elements; and it also teaches how a colored substance may be made to lose its color or change it into such and such another color. Yes, the making of dyes is an important part of chemistry's work, and as the opportunity now occurs to give you some acquaintance with this interesting subject, I am glad to make the most of it. By using an acid we change the blue of the violet to red, and by using lime we change it to green. These two transformations, so complete and so instantaneous, give you an idea of how chemistry, with all the various drugs at its disposal, can produce the many different colors used by the painter and the dyer.

"I again take up my violet, turned green by the action of the lime, and dip it into this glass of water containing a few drops of some acid or other, no matter what, though in this instance it happens to be the acid that results from burning sulphur and is called sulphuric acid. We shall get better acquainted with it later. I could make phosphoric acid serve equally well if you had not used up our supply in your experiments with flowers. Now keep your eyes open for what happens. In the liquid acid the violet turns red, just as it would have done had it not been first subjected to the action of lime. When it is changed to red, I take it out of the glass and place it a second time in contact with lime, whereupon it turns green again. If it were dipped into the acid once more, it would turn red, and then if touched with moist lime a third time it would again change to green. Thus the flower would continue to turn alternately green and red from being alternately subjected to the action of lime and of acid."

"Could these changes from green to red and from red to green go on a good many times?" asked Emile.

"As many times as you pleased. On coming out of the acid the violet would be red, and after touching moist lime it would be green. Oxids of iron, zinc, and magnesium do not have this property of turning violets and other blue flowers green, whereas the oxid of calcium, or lime, does. Whence comes this difference in the properties of these various metals? It comes from the same cause that determines the presence or the absence of taste. Lime, which dissolves in water, acts on the organs of taste, and it also acts on blue flowers by turning them green. The other three oxids, those of iron, zinc, and magnesium, do not dissolve in water, are tasteless, and do not turn blue flowers green, but leave them unchanged.

"But if these oxids were soluble in water, it is probable that they would have a burning taste, more or less like that of lime, and would turn violets and other blue flowers green. Indeed, we know that, besides lime, some other oxids are soluble in water, and all these oxids without exception have the burning taste of lime, but much more pronounced; also, they all turn blue flowers green. Taking for the present no account of any but the soluble compounds, we may sum up in a few words what we have learned about acids and oxids.

"An acid, which is a metalloid combined with oxygen, has a sour taste, and turns blue flowers red. An oxid, which is a metal combined with oxygen, has a burning taste, and turns blue flowers green.

"I must now tell you that an acid and an oxid can combine and thus produce a compound substance, whose properties are, of course, different from those of either the acid or the oxid. You have not forgotten, I trust, that after chemical combination has taken place one must not expect to find in the resulting compound the same properties that were in the separate substances before they united. You have a fairly good notion of phosphoric acid, having seen it made by burning phosphorus and having ascertained its unbearable sourness by tasting it. Lime, a very common substance, is still better known to you, and its burning taste is even at this moment on the tip of your tongue. Now, you would never guess what happens when we combine these two, this acid and this oxid, both of them harmful substances that must needs inspire us with grave mistrust. They are changed into something of the most harmless nature imaginable,—into something, indeed, that is most necessary in the animal body: they become the rigid material that gives strength to our bones.

"Throw the bone from a leg of mutton or from a mutton chop into the fire, and you will see it burn; but what takes fire in this fashion is the grease and other animal matter that impregnate the bone. When this flame has subsided, the bone will be found in its original shape, pure white, and so fragile as to crush under the pressure of your fingers. There we have the essential elements that go to make bone, and they are alone by themselves, purged by fire so as to contain no alien matter. Being incombustible, they have undergone no change by heat, whereas all else has been consumed.

"Now, chemistry tells us that this white, stony material to which the burnt bone has been reduced is formed—not entirely, but nearly so—of phosphoric acid and lime in combination. Grind this hard, white substance to a powder and taste it, and you will find it has no taste, whether sour or caustic. It is as if the substance contained neither phosphoric acid nor lime. It has no effect on either violets or any other blue flowers, their natural color remaining unchanged and without the slightest trace of red or green. In short, the properties of both acid and oxid have disappeared; what was active has become inert, and what was salt and what was caustic no longer have any taste. This stony material of bones, a combination of phosphoric acid and lime, is called phosphate of lime. It contains three elements, phosphorus, calcium, and oxygen, and therefore is, as we say, a *ternary* compound.

"There are innumerable similar compounds,—compounds, that is, made by combining an acid with an oxid. Chemistry calls them salts. Thus, the white, stone-like substance left after a bone is burned, and known as phosphate of lime, is a salt."

"You call that salt, that bone powder that has no salt taste at all?" exclaimed the boys, surprised to hear thus used a word they thought they knew the meaning of.

"Notice, my young friends," their uncle corrected them, "I do not say the hard substance of the bone is salt; I say it is *a* salt. In common language the word 'salt' is used as the name of what we add to our food to heighten its flavor, as when we salt our soup or fish; but chemistry gives to this term a much wider meaning, applying it to every compound resulting from the union of an acid with an oxid. Every acid—and the number of acids is large—can combine with one or another of the oxids, which are still more numerous, and thus are formed salts in great number. Kitchen or common salt has given us a general name which we apply to a great many other compounds; but it is an unfortunate misnomer in its common use, for our ordinary salt really does not belong in the class of salts as I have just defined them, not being composed of an acid and an oxid. Let us, then, disregard the salt we all know so well, let us forget its so-called salt taste and its household use, and henceforth let us understand by salt in chemical language any compound of an acid and an oxid, no matter what its taste, color, or appearance may be.

"All these particulars, as a matter of fact, vary a good deal in the different salts. There are numerous salts that resemble kitchen salt in appearance, being colorless, vitreous (or glass-like), and soluble in water; and from this superficial resemblance to common salt they get their general name. Others are blue, containing oxid of copper; and others green, containing oxid of iron; while others still are yellow, reddish, or violet. In fact, we may find among them almost any color, but not one of these true salts has the peculiar taste of our kitchen salt. Some are bitter, some sour, some caustic, and some of an indescribable taste, but nearly all are intensely disagreeable. There are many, too, that will not dissolve in water, and so are tasteless; such is the hard material of which bones are chiefly made, and such, also, the freestone used for building. To this list we may add the plaster of Paris used in the ceilings of our rooms. Here, then, are three things—the mineral substance of bones, freestone, and plaster of Paris—that you certainly did not suspect of being salts."

"I had no idea of it," replied Emile. "If houses are built of salts such as freestone and plaster, I see that from a chemical standpoint the salt in sausage and ham must be something very different."

"Yes, very different indeed, for what chemists call a salt is to be found in almost any stone on the road, rock on the mountain, or handful of earth in the field."

"Then there must be quantities and quantities of those compounds of acid and oxid."

"Yes, some are very abundant, forming the greater part as they do of rocks and stones and other minerals. The salt called carbonate of lime is one of them. Freestone, rubble-stone, limestone, from which we get lime, and also marble and many other kinds of stone are composed chiefly of carbonate of lime."

"And what is plaster of Paris called in chemistry?"

"Sulphate of lime. But these words mean nothing to you, and so we must stop here and have a little lesson in chemical grammar."

"Does chemistry have a grammar?"

"It has a language of its own, and therefore it has rules for correct speaking when referring to the things it has to do with. But Emile has no need to be frightened at this word 'grammar'; it has nothing to do with his hated conjugations. A few simple rules will make us master of it. Let us begin with the names of the acids. We know that an acid is a burnt metalloid, or, in more precise terms, a metalloid combined with oxygen. That of phosphorus is called phosphoric acid, and this example gives us our rule: to the name of the metalloid, or to the chief part of that name, is added the ending *ic,* and we have the name of the acid. Surely, there 's nothing hard about that. *Phosphorus* and *ic,* with the omission of *us* to make a pleasanter sound, unite and give us *phosphoric,* and that 's all there is to it; we have the name of the acid of phosphorus.

"Let us take another metalloid. You are acquainted with nitrogen. I told you of its unwillingness to combine with oxygen. Nevertheless, by using skill and ingenuity enough, we may overcome this unwillingness and make the two elements unite. What will the acid thus formed be called?"

"According to the rule, I should say nitric acid," answered Emile. "Is that right?"

"Perfectly. Another metalloid is called chlorin, which you know nothing about as yet; but no matter, you can give the name of its acid."

"It must be chloric acid, or I am much mistaken."

"You are not in the least mistaken. Chloric acid is the correct name."

"That 's easy enough. Oh, I wish our grammar at school—"

"Never mind your grammar at school just now, but let us proceed with our grammar of chemistry. From carbon we get an acid that you can name by following the rule already given."

"*Carbon* and the ending *ic* make *carbonic,*" volunteered Jules. "It must be carbonic acid."

"In the same way you will be able to name the acid derived from sulphur."

"Oh, I have it!" cried Emile. "It is sulphuric acid."

"Yes, that is right. But enough, now, on the naming of acids. Let us pass on to that of oxids, which is simpler still. We say oxid of iron, oxid of zinc, oxid of copper, and so on, according to the metal in the compound we wish to name. I need only tell you to notice that certain oxids have retained their common name; familiar to people in general from time immemorial, they have kept in chemistry the name acquired by long use. Thus, oxid of calcium is simply lime, both to the chemist and to the mason. The learned term gives way to the popular one. As we go on we shall meet with other oxids of the same exceptional sort in respect to name.

"There still remain to be considered in this connection the salts. They are formed, as we have already seen, by combining an acid and an oxid; and the rule to be followed in naming them is simple enough. In the name of the acid replace the ending *ic* by *ate,* follow it with the name of the metal supplying the oxid, and you will have the term denoting the salt. Thus nitric acid and oxid of zinc form a salt called nitrate of zinc; carbonic acid and oxid of lead form another called carbonate of lead."

"I see," said Emil; "and so, phosphoric acid and oxid of zinc would make the salt called phosphorate of zinc."

"According to the rule that is quite right, but not right according to usage. The ear is hard to please, even that of a chemist, who, without boggling over the trifles of language, nevertheless finds *phosphorate* a little disagreeable; and so for the sake of euphony the word is shortened to *phosphate,* as phosphate of zinc, phosphate of copper, etc. In the same way *sulphurate* is shortened to *sulphate,* as sulphate of iron, sulphate of lead, and so on.

"One word more and we shall have done with the subject. When a salt contains an oxid that has kept its common name instead of the name of the metal, this common name is still retained. Take as an illustration the salts whose oxid is furnished by calcium. We do not say sulphate of calcium, carbonate of calcium, but sulphate of lime, which is plaster of Paris, and carbonate of lime, which is limestone. Herewith ends our grammar."

"The whole of it?"

"Not the whole, but the most important part."

"Oh, I wish our school grammar—"

"Tush! Don't worry about that."

## CHAPTER XIII

### A TALK ON TOOLS

ON the following day Uncle Paul resumed his talk.

"This gas called oxygen," said he, "the object of our search at present in our exploration of the field of chemistry—have I lost sight of it by allowing so many stages to our journey and thus giving you time to look about and get your bearings? Have we been straying from the direct road? Not by a hair's breadth; we are nearing our goal; in fact, we have reached it. We have just learned that a salt is a storehouse of oxygen, that it is indeed doubly so, for its acid constituent is composed of this gas and a metalloid, while its oxid contains this same gas plus a metal. A salt, then, which unites in itself two burnt substances, is what we must go to if we wish to obtain the gas that makes things burn. Nevertheless we must use care in making our selection, for most salts stoutly resist decomposition and it is only with great difficulty that they can be made to give up their oxygen. We should have no better luck with them than with phosphoric acid or oxid of zinc. What they have once got they hold on to with a tight grip. Consequently, we should be at a loss what to do if chemistry did not tell us the salt to resort to. It advises us to try chlorate of potash as being rich in oxygen and easy to decompose."

A bottle containing a white substance in the form of little transparent scales was set before the boys.

"Here we have some chlorate of potash," their uncle announced. "It came from the drug shop where I bought our stick of phosphorus and some other materials needed in our lessons."

"That looks a little like salt," Emile observed.

"Yes, it looks like it, but its properties are very different. In particular, it has no salt taste; and also it contains oxygen in abundance, while kitchen salt has none at all. I must remind you once more that by an unfortunate accident of language the general term for all salts is taken from a substance that is not, itself, a salt according to the accepted definition; that is, it is not a compound of an acid and an oxid. I will also ask you to bear in mind that many salts have the colorless and glassy appearance of our common salt, and that this likeness, which is purely external, is responsible for the adoption of the term as used in chemistry."

"Then you say this white stuff, this chlorate of potash, has oxygen in it, the gas that makes things burn?"

"Yes, it has oxygen, and a great deal of it, too; so much, in fact, that a little handful of this salt will give us several liters of pure oxygen. It is there, squeezed into a small space by being combined with other things. Go back to our grammar of chemistry, of which you have learned the most important rules, and it will tell you what chlorate of potash is made of."

"The word chlorate," replied Jules, "tells me that the substance contains chloric acid. What that acid is like I don't know, for I have never seen any; but I know at least that according to its name it has in it a metalloid, chlorin, and, besides that, some oxygen."

"Right here let me add," interposed his uncle, "that this metalloid called chlorin is also present in kitchen salt. That will help you to remember the name, which is so new to you, and the time will come before long when you will become better acquainted with the thing itself. What else does the name chlorate of potash tell you?"

"It tells me that the salt contains an oxid furnished by a metal called potash, if I am not mistaken."

"You are mistaken as to the name of the metal, but that is not your fault, for here we meet with an exception like that we found in the use of the word 'lime.' You have not forgotten that certain well-known oxids retain in chemistry their popular names. We say 'lime,' and not 'oxid of calcium'; and in similar manner we say 'potash' instead of 'oxid of potassium.' As to this metal potassium, it is really a metal, much like the metal element in lime, but even softer and quicker to catch fire in water. It is found in wood-ashes. But we need not discuss it further to-day; only notice, while we are on the subject, what curious facts are to be learned from the commonest things when they are examined chemically. Thus, the salt we use in our cabbage soup gives us chlorin, a very interesting metalloid that Emile would not soon forget if he should venture to be too familiar with it; and the ashes from our hearth furnish potassium, a metal that catches fire at the mere touch of water. Briefly stated, potash is the oxid of the metal called potassium, and it is customary to say 'chlorate of potash,' 'sulphate of potash,' and so on, just as we say 'sulphate of lime' instead of 'sulphate of calcium.' So, you see, the salt we have here gets its oxygen both from the acid it contains in the form of chloric acid and from that present in its oxid, potash.

"This salt decomposes readily, needing only heat to make it surrender all its oxygen. Look here a moment, and you will be convinced."

So saying, Uncle Paul dropped a pinch of chlorate of potash upon a handful of glowing coals, whereupon the chlorate of potash melted with much bubbling and foaming, while the coals all about it began to burn with extreme brightness and heat. It was as if a puff of wind had fanned the glowing embers, though no puff of wind could have been at once so noiseless and so effective. The intense burning went on with an energy astonishing to the young beholders.

"What splendid kindling that would make," Emile remarked, "for starting up a fire when it gets low! All you 'd have to do would be to throw on a handful, and the wood or coal would be all ablaze.

You might work the bellows all day and not make it burn so."

"From the bellows," replied Uncle Paul, "comes only air, in which nitrogen, a gas that does not help things to burn, is much more plentiful than oxygen, which alone helps on the burning. In this mixture the inert or useless gas considerably weakens the effect of the active or useful gas. But chlorate of potash, on being decomposed by heat, sends out a breeze, as we might say, of pure oxygen; and that is why the coals here burned just now with so much intensity. With this kindling, as Emile calls it, a fire gains in vigor and rapidity because the gas it feeds upon is mixed with nothing to weaken its effect."

More pinches of the chlorate were thrown upon the coals, and when the two young observers had seen how this easily melted substance quickened the fire by the release of oxygen, which escaped in bubbles, Jules told his uncle of something he and his brother had once noticed with much curiosity.

"One day," said he, "I had taken a feather and swept off a handful of that white, moldy stuff that collects on cellar walls, when some one told me it was saltpeter and that saltpeter is used for making gunpowder. I dropped some on the live coals in the fireplace, and it made a quick, bright fire like that you made with the chlorate of potash. Does that fluffy stuff on damp walls give out oxygen when you put it on the fire?"

"The fine white flakes such as you swept from

the cellar walls are, in truth, saltpeter,—or, in chemical language, nitrate of potash. This substance is, as its name shows you, a salt, containing oxygen in its acid part, which is nitric acid, and also in its oxid, which is potash. When you throw it on the fire it decomposes, releasing its oxygen; and that explains why it makes any burning substance burn so much faster than before. Thus, the saltpeter from damp walls acts in the same way as chlorate of potash; they both decompose, and in so doing give out in abundance the gas needed to make fire burn. I must let you know, however, that nitrate of potash is not a suitable substance from which to obtain oxygen, because it is not so readily decomposed as Jules's experiment would seem to indicate. To make this nitrate surrender its oxygen heat is not enough; it also requires some combustible matter, as wood or charcoal. Then the fuel seizes upon the oxygen as fast as it is released, and so the gas escapes us again, being taken and held captive in another compound. Consequently, nothing has been acomplished; what we wished to obtain has slipped out of our grasp and entered into a new combination. With chlorate of potash, on the contrary, heat alone is enough for our purpose, without the help of anything else that might appropriate the oxygen releasd by the salt."

"Another question," said Jules.

"As many as you like, my young friend. I shall take pleasure in answering them, for I know beforehand that they will be well-considered questions, coming as they do from a thoughtful mind."

"When you threw the chlorate of potash on the coals, it first melted, then bubbled and let go its oxygen, and at last there was nothing left of it but a little white round piece that would n't burn at all. What is that white stuff that is left on the coals?"

"Your question is well put, for it concerns a rather important matter. I had forgotten to explain this, but now I will repair the omission. This remainder, this bit of white crust that fire has no effect upon, comes from the chlorate of potash when the latter is decomposed by heat. What did this chlorate contain to start with? Three elements, the chlorin in the chloric acid, the potassium in the oxid of potassium, and the oxygen in both. Of these three elements, one, oxygen, has disappeared. There remain then, the chlorin and the potassium, united in a compound very different from the original chlorate. This compound is called chlorid of potassium.

"This gives me an opportunity to teach you a new rule in the grammar of chemistry. The various metalloids can combine, one by one, with the various metals. In the case of oxygen, for example, the combination with a metal is called an oxid, as you already know; and so of the other metalloids, sulphur, chlorin, phosphorus, etc., the name of the compound is formed by adding the ending *id* to the name, or the chief part of the name, of the metalloid, after which comes the name of the metal preceded by the preposition *of*. In the substance before us we have chlorin and potassium in combination; in other words, we have chlorid of potassium.

But enough of this, and perhaps too much; let us return to the subject of oxygen.

"We have to find out, if possible, how an unskilled experimenter may hope to obtain, without too much difficulty, the gas stored up in the chlorate of potash. He should first procure a glass receiver of some sort, and in this the decomposition is to take place. A medicine-bottle, short and as wide as possible, will answer the purpose, provided the glass is not only thin, but of uniform thinness throughout. A glass vessel exposed to heat will remain unbroken only if it complies with these requirements. The thinner it is, the less likely is it to break under sudden variations of temperature. Look at this tumbler here: at the bottom it is as thick as your finger, but it is thin elsewhere. Plunged cold into hot water, or hot into cold water, it would run great risk of breaking. On the other hand, a piece of glass of even thinness comes out intact from a similar test. Let us, then, choose the thinnest bottle we can find, and above all one that has no thick places in the glass as are often found in the bottom of an ordinary bottle. The success of our experiment depends greatly on our choice."

"I should have thought," said Emile, "a thick, strong glass would be the kind to use. It would stand the strain better."

"Yes, if it were a question of resisting shock or of not melting under heat. But here it is not a question of resistance to shock, as I take for granted the operator is sufficiently skilled not to bump his receiver against any hard object; nor is the danger of melting worthy of consideration, the heat required for decomposing the chlorate being insufficient to fuse or even soften the glass. Our bottle, then, will not have to withstand shocks, and it will not have to bear any excessive heat, though it will undoubtedly undergo changes of temperature sudden enough to break it unless we guard against such an accident by selecting a bottle of very thin glass."

"But what if, after all, the bottle full of chlorate should break over the fire; what would happen then?"

"Nothing very serious. We should simply have to let it alone and watch the beautiful fireworks resulting from the release of so much oxygen on burning coals. We should witness on a large scale what you saw when I put a few pinches of the chlorate on a handful of glowing embers."

"And then?"

"Then we should begin again with another bottle; that 's all. But better than a medicine bottle, which I should unhesitatingly use if I had nothing more suitable, is a receiver called in chemistry a balloon. It is a vessel of clear glass, round in shape, with a neck about as long as your hand is broad, and it can be had for a few cents at the druggist's, or some chemist might spare us one from his laboratory. Here is the balloon we are going to use; it is one of my recent purchases in town."

"That looks like one of those candy bottles they always have for selling candy to the children at village fairs. For two cents you can buy the whole thing, candy and bottle and all."

"Your candy bottle would do very well here if it were larger; but what could not well be replaced by anything else is the tube for conveying the gas from the balloon into the bell-glass. The tube is made of glass, and may be had of the druggist, who will perhaps furnish the complete outfit we need, all ready for use, or we might apply to some chemist friend. But, after all, our best course is to make our apparatus for ourselves, and then we shall be sure to have it in all respects as it should be for the successful performance of our experiment. At the druggist's we find straight glass tubes of a meter and more in length. We select several about as large round as a lead pencil and made of thin and colorless glass, which softens much more readily with heat than does greenish glass. Let us examine the thickness of the glass. If in cross-section it looks thin and colorless, it is what we, with our limited resources for working in glass, are seeking. A thick green glass would be too unmanageable for us. Providing ourselves, therefore, with a few of these straight tubes of thin, clear glass, we next proceed as follows:

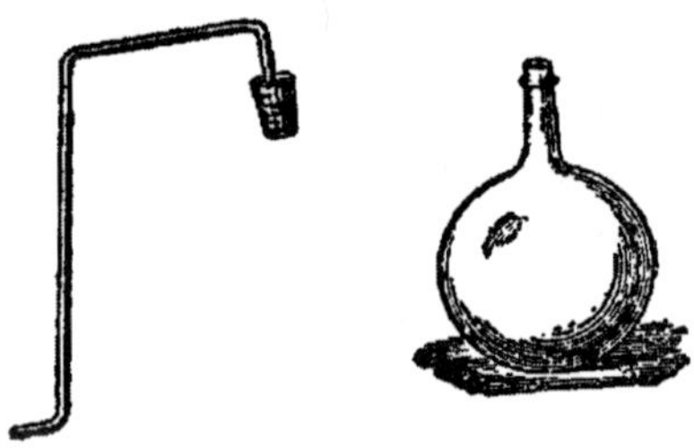

"To cut from the straight tube any desired length, we first take a triangular file and make a little groove all around the tube at the point where it is to be broken, then, taking the tube in both hands, we hold the grooved part against the edge of a table and exert a gentle pressure. The break is made instantly, round and clean. Now it remains to fashion this broken-off piece so as to give it the shape suitable for our experiment. All we have to do is to bend it at various points after first softening it at those points by heat. With easily melted glass this may be done with the help of a handful of live coals blown to glowing heat by the breath; but the operation goes better if we use an alcohol lamp. This is simply a metal or glass cup or holder containing alcohol into which runs a wide cotton wick, the exposed end of which is lighted. The part of the tube to be softened is held to the flame by taking one end of the tube in each hand and turning it between the fingers so that the heat may get at it all round. As soon as the glass seems soft enough to bend, a slight exertion of strength will produce the elbow desired, which may then be left to cool off slowly.

"The tube thus bent is to be attached to the balloon by means of a stopper with a hole through it, the stopper being one that fits tightly and allows no gas to escape. This perfect stoppage is necessary in handling oxygen and other gases, so subtle is their nature. The smallest possible opening is big enough for them to get through. Hence the stopper must be fitted with extreme nicety. I will explain how to do it.

"Select a stopper of fine cork, as uniform in its structure as possible, and without any of those holes or decayed spots that are found in cork of poor quality. With some heavy object, the first to come to hand, such as a round stone or a hammer or anything else of the sort, first give the stopper a few light blows to soften it and make it supple. Then with a coarse iron wire pointed at one end, and heated red-hot if you choose, for the sake of easier and quicker execution, pierce the stopper lengthwise, making a small hole for the guidance of the file with which it is to be enlarged. This latter, called a rat-tailed file from its shape, is round and should not be of greater caliber than the tube that is to pass through the cork. With this file you then carefully enlarge, perfect, and smooth the channel in the cork until the tube will pass through under gentle pressure and fill it exactly. Now take your cork in hand again and shape the outside to fit tightly into the neck of the balloon. With a coarse flat file you first rasp it lengthwise to make it regular in shape and slightly tapering. When thus reduced to the proper size, it is finished off with a finer file, being worn down and smoothed until it fits perfectly into the neck of the balloon. Observe that no knife or other edged tool, however sharp, could do the work of the file in thus preparing the stopper, for the cork would be cut unevenly and the result would be an escape of gas. An accurately fitting stopper is indispensable to success. In future, then, we will count as necessary to our laboratory equipment four files,—namely, a fine triangular one for notching the glass tube so that any desired length may be broken off;

a round one for enlarging the hole made in the cork by the iron wire; a coarse flat one for rasping the outside of our stopper and giving it a preliminary shaping; and, lastly, a fine flat one for finishing and smoothing."

During these explanations the speaker's hand helped out his words, example accompanied precept, and the tube was bent in the flame of the alcohol lamp, the stopper was pierced and filed and finished, and in a short time everything was in readiness.

"Our apparatus is now in order," said he, "and we will put it to use. But first a brief word of explanation will not be out of place. Heat alone suffices to decompose chlorate of potash and make it give up its oxygen, although toward the end of the operation the salt becomes obstinate and, in order to obtain all the gas, must be heated to such a temperature as would endanger our balloon by softening the glass. I hold it of the first importance, and you will agree with me, that we should not damage our apparatus, as our resources do not admit of our sacrificing a balloon every time we want a little oxygen. Besides, it is much more convenient to operate with less heat. Well, then, chemistry tells us that by mixing the chlorate with some black substance that will distribute the heat evenly throughout the whole mass the complete decomposition of the salt may be easily accomplished. A handful of live coals will under these conditions give enough heat, and the balloon will not be exposed to any danger.

"My speaking just now of a black substance perhaps made you think of coal-dust. Beware, my dear boys, of mixing anything of that sort with the chlorate and then heating the mixture, if you do not wish to be disfigured by a violent explosion. Those two substances make a dangerous mixture. And why? The reason is plain. On being heated, the chlorate gives out oxygen, and this gas, mingling with an inflammable powder such as coal-dust, will not fail to produce a sudden explosion that would blow our apparatus to pieces. Nothing that can burn should be mixed with the chlorate, the risk involved being too great. Bear this well in mind.

"What, then, shall be the black powder for making quicker and easier the decomposition of our chlorate? It must be something that will not burn; something that, having already been burned, that is united with oxygen, will no longer catch fire. The best thing for our purpose is a metal oxid. There is found in certain mines—and the druggist sells it for a trifle—an intensely black powder called dioxid of manganese. Manganese itself is a metal much like iron, rarely found in a pure state, and hardly ever used in that form. United with oxygen, it forms various compounds, the second of which in abundance of oxygen is the dioxid I have mentioned, which is what we require for mixing with our chlorate to help on its decomposition. With this black substance we incur no risk; fed up, so to speak, with the gas that makes things burn, it can take on no more, a fact that removes all danger of any excessive combustion within our apparatus.

"Accordingly, I place on a sheet of paper a good-sized handful of chlorate of potash, a little less of dioxid of manganese, mix the two together, and pour the mixture into a balloon of the size of a large orange. The tube with its stopper is then adjusted, and the apparatus, supported by a triangle of stout iron wire, is placed on a brazier.

"Here will occur a slight difficulty that must be overcome before we can proceed. Our oxygen is to be collected in wide-mouthed bottles or jars, full of water and inverted in the bowl of water so as to bring the free end of our glass tube directly under the mouth of the inverted bottle or jar, and the latter must accordingly be held in a tilted position. But this is tiring to the arm if the operation is of any length, and it would be better if the inverted bottle could be made to stand erect on some support. But how, in such a position, can a passage be left open for the end of the tube conveying the gas from the balloon? Nothing simpler. Let us take a small flower-pot having a hole, as is usual, in the bottom; then by breaking off some of the upper part of the pot we can lessen its height and obtain a sort of cup only a few fingers tall. No matter if the edge of this cup is irregular or even jagged; it will do if it stands firm and if its upturned bottom offers a level support for the inverted bottle or glass jar. Finally we make a deep notch in the side, and the support for our bottle is ready. We place it in the middle of the bowl, the flat bottom upward, and through the notch in its side we pass the end of our glass tube, which will thus emit the released oxygen within the enclosure of the supporting pot. On the latter will rest the inverted bottle or jar full of water, and the gas will gain access to it through the round hole in the support.

"Enough, my little friends, on this subject. Our apparatus is harder to explain than to make. I promise you for to-morrow some experiments that will richly repay you for the dry preparations of to-day. And now get me one more sparrow, if you please. Set your snares in the bed of peas. As the captive bird will not share the sad fate of its predecessor, there is nothing to dread in our coming experiment."

## CHAPTER XIV

### OXYGEN

AT last they were about to see this famous gas, oxygen, this king of elements that in the past few days had again and again been referred to by Uncle Paul in his talks, but always wrapped in the mystery that enshrouds the unknown. Now it was to be released from its imprisonment in the chlorate of potash and subjected to tests that promised to be interesting. Emile even dreamed about it in the night, so much had his mind dwelt on this gas that makes things burn. In his dreams he saw the glass balloon and the bent tube cutting up all sorts of silly capers on the circular edge of the brazier, while the irascible chlorate and its comrade, the dioxid, looked on curiously through the glass walls of their prison. In the presence of the real things about which he had dreamed the night before—confronting, that is, the various preparations for the coming experiment—the boy was moved to fresh laughter when his uncle put the balloon in place over the live coals.

They had not long to wait. Before it seemed possible that the heat could have had time to act, and with no visible change in the contents of the balloon, the water in the bowl began to bubble at the end of the tube, a sure sign that gas was being set free. The support already made from the bottom of a flower-pot was arranged in position in the bowl of water, and a large bottle of two or three liters' capacity and with a wide mouth was filled brimming full of water, stopped with the palm of the hand, and turned upside down on the support. Emile steadied it with his hand. The gas passed through the hole in the middle of the support, and rose through the water in a tumultuous and uninterrupted series of big bubbles, so that in a few moments the bottle was filled with the gas that had

displaced the water. Then Uncle Paul took a tumbler, plunged it into the bowl, and in this glass of water set the mouth of the bottle, of course without letting it come out into the air and lose any of the imprisoned oxygen. That done, the bottle, standing upside down in the glass of water, which served to exclude the air from outside, was set safely away in a corner of the room to wait until it was needed for some experiment still to come. A second bottle took its place in the bowl, was filled with gas in the same way, and set aside for future use, after which a third and then a fourth went through the same operation. The supply of oxygen in the balloon seemed inexhaustible.

"There seems to be a lot of oxygen in a handful

of chlorate," remarked Emile, surprised at the quantity of gas being released.

"Yes, it is no small amount, as you see, for our four bottles hold almost a dozen liters, taken together."

"And those dozen liters of oxygen were all in that little heap of chlorate?"

"All were in the small amount of chlorate used. Was n't I right in calling this salt a rich storehouse? The chlorate is more than abundantly fed with oxygen; it is crammed to the limit. The gas taken captive by chemical combination is compressed in great quantities into very small bulk. But all is not over yet; I hope to fill this bottle, too."

Therewith Uncle Paul placed on the support in the bowl an odd-looking, elongated bottle that had hardly any neck, being of nearly the same width at top and bottom. He did not say where it came from, but his nephews thought they recognized it as an old pickle-jar. They were smiling at seeing this homely utensil pressed into such serious service, when their uncle resumed his talk.

"Does my tall jar make you laugh? Because it once held pickles, do you think it unworthy now to hold oxygen? Away with such false pride, my lads! Let us use the things at hand and not sigh for expensive luxuries. But for this once we will carry out our program in a worthy manner, just as we should if we had a better-equipped laboratory at our disposal.

"Here is something that chemists call a gage It is a tall glass cylinder with a base to stand on.

I proceed to fill it with oxygen while there is still some left in the balloon, and we see that the gas enters slowly, the supply being nearly exhausted. In the balloon, the contents of which show no change in appearance, there is left the dioxid of manganese just as it was when I put it there. It has suffered neither loss nor gain, but it has promoted the decomposition of the chlorate by enabling the heat to act upon it evenly. The balloon itself, too, remains uninjured and ready for further use when needed. As to the chlorate, it has now lost all its oxygen, and is thus changed to that white substance we saw yesterday remaining on the embers after we had thrown on a pinch of chlorate and watched it quicken the fire. In short, it has turned into chlorid of potassium. So much for that. Now let us put our supply of oxygen to some use, beginning with that in the gage."

Carried from the bowl to the table with the usual precautions—that is, being first closed with the palm of the hand while still standing upside down in the water—the gage was set upright on its base and covered with a piece of glass pending certain preparations consisting of fastening a short piece of candle to an iron wire, as had been done before in the experiment with nitrogen. Uncle Paul lighted the candle, allowed the flame to get bright and full, and then blew it out; but the wick was still glowing from the combustion that had been so suddenly arrested.

"The candle which I have just blown out," said he, "but which still shows a red glow at the end of the wick, I am going to lower into the oxygen in the gage. What will happen? We shall see."

Removing the piece of glass, he suited the action to the word. *Piff!* A slight explosion was heard and the candle, relighting itself unaided, burned with radiant brilliance. It was extinguished once more, the wick still retaining its glowing spark, and lowered again into the gage of oxygen. Another *piff!* and the flame reappeared, burning with remarkable brightness. Again and again, with care to leave the wick slightly aglow, the candle, extinguished with a breath, quickly relighted itself when it was lowered into the oxygen. Each time a little explosion preceded the revival of the flame. Emile clapped his hands with delight at this repeated renewal of the flame, always so prompt and so complete.

"How different it was with nitrogen, the partner of oxygen in the air we breathe," he observed. "Oxygen relights all of a sudden what is just about to stop burning altogether, but nitrogen puts out what is already well on fire. Could n't I try my hand at this fine experiment, Uncle?"

"Surely; why not? But I must inform you that the oxygen in the gage is by this time nearly exhausted, the candle having used up a little every time it was relighted."

"But there 's plenty more in those four bottles."

"I am keeping that for other experiments of still greater importance."

"What shall I do, then?"

"You must content yourself with my old pickle-jar that I took the pains to fill with oxygen, counting on it to take the place of a regular gage."

"I 'll do that and be glad to."

"A wise decision, for the old pickle-jar will render you yeoman service. My chief reason for using it is to show you, by adding example to example, that instructive experiments are possible even with the commonest utensils. Our gage here is a luxury, an unheard-of extravagance in our little village. Almost any kind of bottle, any caper- or pickle-jar, provided only it has a wide mouth to admit the candle, would serve very well in performing the striking experiment you now wish to repeat. All right, then; you shall repeat it."

The jar being placed on the table, Emile began the lighting and relighting of the candle, blowing it out and rekindling the flame again and again. The experiment had not gone better even with the regular gage.

"There, now," said his uncle, "does n't my pickle-jar answer the purpose admirably?"

"Yes, splendidly."

"It is the contents and not the container that we should give our attention to. If we supply it with oxygen the candle will relight itself, no matter what the oxygen is in, whether the chemist's gage or the paltry pickle-bottle. As a close to the experiment, leave the candle in the jar to burn as long at it can. You will see how fast it will be used up."

And, indeed, the candle, immersed in oxygen, did not fail to burn with devouring rapidity. It showed no longer the calm flame that is maintained in ordinary air, but a furious tongue of fire, extraordinarily bright and excessively hot, making the wax melt and run down in big drops. The substance of the candle was literally devoured rather than burned, and it was evident that in a few minutes there would be consumed in this energetic gas enough wax to last an hour in atmospheric air. Finally the flame died down for lack of oxygen, and Uncle Paul resumed his talk.

"Before continuing these spectacular experiments with oxygen," said he, "let us quiet our emotions with a little interlude. You know the characteristics that enable us to recognize an acid,—first the sour taste, and then the property of turning blue flowers red. But it is not always practicable to test an acid by the sense of taste, the flavor being sometimes very weak or even quite imperceptible. The test with blue flowers is a better one. But, unfortunately, violets and other blue flowers redden with some difficulty when the acid is weak. Chemistry has found the blue coloring matter of lichens more easily affected. You know those peculiar growths that look like flaky crusts on the bark of trees and even on the surface of the hardest rocks. They are of vegetable origin, and are called lichens. One species found on rocks near the sea furnishes a blue substance called litmus. The druggist has this for sale in the form of little cubes, ashy-blue in color and known as litmus tablets. If you dissolve one of these in a little water, you obtain a pale violet-blue liquid called tincture of litmus.

"This tincture is one of the most convenient

tests[1] for applying to acids, as it reddens much more easily than do blue flowers. To illustrate, I pour into this glass two fingers of the litmus tincture; then I dip the end of a glass tube or of a common straw into the acid in this bottle, an acid I have already referred to as derived from sulphur,—sulphuric acid, in fact. I do not dip it deeply, but merely touch it to the liquid, and with this barely moistened straw I stir the blue tincture, which immediately turns red in proof (if I did not already know it) that what my bottle contains is an acid.''

''If the tincture of litmus is turned red by an acid,'' said Jules, ''it ought to be turned green by a soluble oxide, just as violets are; and so it would help us to find out whether a thing was an oxid or not.''

''It is perfectly natural to expect that, after seeing what happens to blue flowers; and yet it is not the case. Lime and other soluble oxids do not turn the blue of litmus green, but leave it unchanged However, this lack is atoned for by another characteristic. Once reddened by an acid, litmus is turned blue again by means of a soluble oxid. Into the contents of this glass, just now changed from blue to red by the addition of sulphuric acid, I drop a tiny particle of lime, and the liquid returns to its original blue. A second time I apply a trace of the acid with the end of this straw, and a second time the tincture turns red. Lime being

[1] Soft paper tinged with litmus and called litmus-paper is now more conveniently used for test purposes.—*Translator.*

applied in its turn once more, the blue color reappears. These changes from blue to red and from red to blue might be repeated indefinitely. Here we have, then, a perfect test for determining whether a substance is an acid or an oxid, provided only, of course, that it is soluble in water. Whatever reddens the blue tincture of litmus is an acid, and what restores its blue to the tincture previously reddened by acid is an oxid.

"If we had no litmus—and the lack of it would be no great matter—we should have to be contented with blue flowers. A bunch of violets would first be crushed and stirred in water, and the bluish liquid thus obtained would be strained and set aside to serve the purpose of litmus. But it would show a difference in one respect; though acids would turn it red, oxids would not turn it blue again, but green; and of course the blue liquid would turn green immediately when acted upon by an oxid, without having to be previously reddened by an acid. It is to be noted, further, that weak acids might not be able to change the blue of violets to red, and therefore litmus is preferable as a test.

"Our interlude is over and we will go on with the performance. We are going to burn several substances in oxygen and watch their manner of burning. First comes sulphur.

"Adopting the method you saw employed when we tried in vain to make phosphorus and sulphur burn in a bottle filled with nitrogen, I make a little cup out of a bit of broken earthenware, and bend the end of an iron wire into a circle for holding this cup. The iron wire is then passed through a large cork stopper that will serve to hold it in place in the bottle rather than to stop up the bottle itself. Hence it does not matter much if the cork is too large for the bottle. A small disk of stout cardboard laid over the mouth of the bottle would answer just as well. The end of the wire, projecting above the supporting cork or cardboard, will serve as a handle for lowering or raising the cup so as to bring it into or near the center of the bottle, in the midst of the supply of oxygen."

Having finished these preparations, Uncle Paul carefully took up one of the large bottles in reserve, together with the glass full of water in which it stood and which served to close its mouth. These, without disturbing their relative position, he carried to the bowl, and there, under water, the glass was removed and replaced by the palm of the hand applied to the bottle's mouth. In this way it was possible to set the bottle upright on the table without bringing its contents into communication with the outside air. A small sheet of glass laid over the mouth closed the bottle, serving as it had done before as a temporary stopper. By means of the iron wire passing through the cork stopper, the cup, which had previously been filled with small fragments of sulphur, was so adjusted that it would take its proper place in the bottle when all was ready for it. Then Uncle Paul set fire to the sulphur and lowered the cup containing it into the oxygen. Thus suspended in the middle of the bottle and held there by its cork support, the cup of sulphur required no further attention on the part of the experimenter, and there was nothing more to do but watch the result.

Every one knows how slowly and with what a dim light sulphur burns under ordinary conditions. Hence the novelty of the spectacle now presented to the astonished gaze of the two young chemists. At their uncle's bidding the shutters had been closed, so that no daylight should get in and dim the splendor of the burning sulphur. It burned with an ardor unapproached by any brimstone match that was ever made. A fantastic radiance of a beautiful violet blue, rivaling in purity the rainbow's purple stripe, emanated from this wonderful illuminant and filled the room with so strange a glow that one might have fancied oneself transported to some other world where the sun is blue.

"Magnificent, magnificent!" cried Emile, clapping his hands with enthusiasm.

The fumes of burning sulphur, escaping in puffs from the bottle and almost suffocating in their intensely pungent odor, tended somewhat to spoil this fairy-like illumination, otherwise so perfect; and so Uncle Paul, as soon as the flame began to die down, had the shutters and windows opened.

"It is all over," said he; "the sulphur has used up its supply of oxygen. I will not dwell on the splendors you have just witnessed, as your eyes have done better justice to them than could any words of mine. They have told you that sulphur burns in oxygen with a heat and a brilliance that it does not have when burning in ordinary air. I will pass on to inquire what has become of the sulphur we have just seen burning so brightly. What has resulted from its combination with oxygen? The result is an invisible gas with a pungent odor, a gas that makes one cough,—the same gas, in fact, that comes from a lighted match. A little of it has escaped into the room—our sense of smell and our coughing tell us that—but a good deal is still left in the bottle. Let us consult our tincture of litmus and see what information it will give us. I pour a little into the bottle and shake it up, whereupon the blue color immediately turns red. What does the litmus say?"

"It says that the sulphur has turned into an acid by burning," replied Jules.

"And it 's a good thing it does say so," put in Emile, "for it would n't have been very pleasant to taste it; and, besides, you can't even see it. That litmus is certainly a convenient thing to have around."

"Very convenient," assented his uncle. "Here is something that can be neither felt nor seen, and yet a very real thing that takes you by the throat and makes you cough worse than if you had the whooping-cough. We wish to know what it is, and our litmus on being consulted answers: 'It is an acid.' "

"Does it say, too, that it is sour?"

"Undoubtedly. What reddens litmus and blue flowers, is always sour."

"But how make sure that the litmus and the

violets tell the truth? I can't stick my tongue down into the middle of the bottle."

"The invisible gas from the burnt sulphur will mix with water, and so a good deal of it is absorbed in the tincture of litmus that I shook up in the bottle. We know this to be so from the effect it has on the tincture, which by itself is only water colored by a tiny particle of matter that has no flavor. Let us taste this liquid reddened by the intermixture of the gas, and we shall find out how our invisible gaseous compound tastes. Wet your finger freely in it without fear of too large a dose. It takes a good deal to make any impression on the tongue."

With their uncle to set them an example the boys tasted the liquid several times to make sure they had its flavor.

"Weak vinegar," pronounced Emile, smacking his lips; "very weak vinegar."

"Weak, if you like, but still vinegar,—that is to say, acid."

"It 's nothing like so strong as phosphoric acid; that eats away your very flesh."

"Our sense of taste, then, agreeing with the tincture of litmus, tells us that sulphur by combining with oxygen in burning becomes an acid. It is this invisible gas, therefore, that has that pungent odor and makes us cough; and it is called sulphurous acid."

"You told us," said Jules, "of another acid made from sulphur, sulphuric acid, which you just used to turn the tincture red. Then sulphur makes two acids, does it?"

"Yes, my boy, sulphur makes two acids, one with less oxygen and one with more. The one with less oxygen, and so the weaker, the less sour, is sulphurous acid; the other, richer in oxygen and therefore stronger and sourer, is sulphuric acid. By simply burning, either in ordinary air or in pure oxygen, sulphur takes on a certain amount of oxygen and no more, being thus changed to sulphurous acid; but by roundabout methods known to chemistry it can be made to take a larger dose, and in doing so it becomes sulphuric acid. Enough about sulphur. Let us next see what will be the result of burning charcoal in oxygen."

A piece of charcoal no bigger than one's little finger was fastened to one end of an iron wire and the other end passed through a small disk of cardboard that was to rest on the mouth of the bottle of oxygen. Uncle Paul then lighted the charcoal in the flame of a candle, but only enough to make it glow at one little point, and in this condition he lowered it into a fresh bottle of oxygen, conducting the operation in the same manner as with the sulphur.

The spectacle that followed rivaled in beauty the one just applauded so enthusiastically by Emile. At the point kindled by the candle, a spark so faint as to be hardly visible, a flame burst forth,—bright, ardent, irresistible,—and, spreading rapidly through the charcoal, soon turned it into a dazzling little forge. It gave an intensely white light, with little sparks snapping and darting in all directions, like so many shooting stars shut up in the bottle. It had taken but an instant to set the charcoal all aglow in a way impossible with any draft of ordinary air. Without moving his eyes from this brilliant spectacle, Emile gave utterance to his thoughts:

"This heat and this bright light and these sparks I can make come when I blow with the bellows on burning charcoal. At the spot just under the nozzle of the bellows the charcoal burns almost as brightly as this, here, in the bottle."

"That is quite natural," rejoined his uncle. "With the bellows you send out air,—that is to say, oxygen mixed with a good deal of nitrogen, this latter weakening the effect. But by a sufficiently rapid renewal of this air that helps things to burn, the glowing charcoal may be made to brighten up and look much like this before us in the bottle of pure oxygen."

The supply of oxygen being at last used up, the unconsumed charcoal turned dimmer and dimmer, and then quite black. The shutters, which had again been closed, were reopened to let in the daylight, the previous admission of which would have greatly lessened the effect of the spectacle.

"What has become of the charcoal that was burnt up? That is the problem we must now solve," said Uncle Paul. "There is left in the bottle an invisible gas having scarcely any odor; and if we trusted only our smell and eyesight, we might conclude that the contents of our bottle had not changed in the least. But let us submit these contents to various tests more decisive than those of smell and sight, and we shall find that there has been a decided change. First, it is safe to predict that, if the charcoal which burned so brilliantly in the beginning will now no longer burn in the bottle, a lighted candle will not burn there any better. Watch. I lower this candle, well alight, into the bottle, and it is hardly inside the neck before it goes out. Consequently, there can be no more oxygen left, for if there were, the candle, as you know, would burn with a bright flame.

"Still another test: I pour into the bottle a little tincture of litmus and shake it up thoroughly so as to give the gas a chance to act on the liquid. The blue tincture changes its color to a very pale red. Shaken up with oxygen, the same tincture would not show the slightest change. Hence, we have here another acid made by burning something in oxygen. We are now convinced that the oxygen in the bottle has been converted into another gas, no less clear, no less invisible, but endowed with very different properties; and this difference, it is plain enough, can be due only to the addition of charcoal (or carbon, which is virtually the same thing) to the oxygen. Hence we must conclude that in this gas, here in the bottle,—a gas so colorless that we cannot see it,—there is at least some slight quantity of carbon, that substance known to us as hard and heavy when we see it in coal."

"I see now that it must be so," assented Emile; "but if any one had told me, without being able to prove it, that there was carbon in a gas as invisible as air, I should n't have been in a hurry to believe him. What do you say, Jules?"

"I say that it is hard to get used to the idea that a thing we can't see or feel can have carbon in it. If Uncle Paul, instead of leading us step by step to where we are now, had begun by saying that there is carbon in this bottle in which I can see nothing at all, we should have looked at him in the greatest astonishment. But the proofs are there and can't be got rid of. The charcoal in burning has changed to a gas that turns the litmus tincture red, and so it must be an acid called— Uncle has n't told us yet what it is called."

"Consult your grammar of chemistry, and find the name for yourselves."

"That 's so! I 'd forgotten all about that. Charcoal is the same as carbon, and the ending *ic* added to *carbon* makes *carbonic*. The gas that comes from burning charcoal is carbonic acid."

"Is it sour like the others, this carbonic acid?" asked Emile.

"Of course, but so slightly that in the gas we have here in our bottle this acid quality is barely perceptible. The litmus, instead of turning a decided red, took on only a faint winy tinge, and to the sense of taste the sourness would be correspondingly feeble. But some day an opportunity will occur to convince you that carbonic acid is indeed sour to the taste. Now let us put our third bottle of oxygen to its intended use. In it I propose to burn some iron, a thing that Emile was inclined to think impossible the other day. And I shall make this iron burn without having to heat it red-hot beforehand in a forge, as the blacksmith does with his iron before hammering it into shape. I shall set it on fire with a piece of lighted tinder, as if it were a train of gunpowder."

"And the iron will catch fire just from that piece of tinder?" was Emile's wondering inquiry.

"Certainly; gunpowder could n't do it better. Here is an old watchspring, broken at one end and of no further use. It is a bit of refuse I got at the clockmaker's. But instead of this spring, which is admirably suited to the requirements of our next experiment on account of its thin flat ribbon-shape, offering plenty of surface to the action of the oxygen, we could use an iron wire about as fine as a medium-sized needle, first cleaning it well with a file or, better, with sandpaper. The watchspring, however, is to be preferred. I begin by heating it over some live coals in order to take away its stiffness and make it supple. Then I wind it around a slender rod of some sort—a penholder or a lead-pencil will do—to give it a spiral or corkscrew form. Next I take a pair of stout scissors and cut one end of our corkscrew into a point, which I stick through a bit of tinder about as large as your finger-nail. Finally, the other end of the corkscrew is passed through a small disk of cardboard which will be placed over the bottle's mouth and will hold the metal ribbon in position in the midst of the oxygen. The spiral of our corkscrew should be pulled out to such a length as to bring its lower end into the middle of the bottle. If a wire is used instead of a watchspring, the same mode of operation should be followed,—the same winding of the wire into a corkscrew form, the wire being first heated if necessary and also (an indispensable detail) cleaned with sandpaper, and the same use of tinder attached to the lower end of the spiral."

All these arrangements being carefully completed, the third bottle of oxygen was set on the table. In preparing it, Uncle Paul had taken care not to fill this bottle entirely with the gas, but to leave several inches of water in the bottom.

"There 's some water in the bottle," Emile pointed out, resolved not to let any detail of this curious experiment escape him.

"Yes, and it is there for a purpose. If there were none, some would have to be poured in now. A considerable depth of water at the bottom is necessary if we wish to keep our bottle for future experiments. You will very soon see why the water is needed. Close the shutters and I will begin."

As soon as the room had been darkened, the tinder was lighted and the spiral ribbon lowered into the oxygen. The tinder flared up suddenly and burned with a bright flame. Next followed a moment of indecision; the iron was taking fire; then it was well on fire and presented the appearance of a piece of fireworks. This marvelous flame, feeding on metal, was seen to make its way upward in a spiral curve as a fire spreads from bottom to top of a winding staircase. Snappings and cracklings and sprays of sparks accompanied the process. At the end of the ribbon there collected and hung suspended a globule of molten metal of dazzling brightness. Growing too heavy, it detached itself and fell. Still intensely hot it plunged through the water with a sharp hissing sound and in a state of redness reached the bottom of the bottle, which it softened as it flattened itself out there. Other globules followed, dropping one by one from the flaming spiral, and despite the cooling effect of the water the largest ones retained enough heat to melt the glass a little and sink into it.

The boys stood silent before this magic spectacle of iron devoured by oxygen; but Emile was not without his fears. The hissing when the melted globules fell, the failure of the water to extinquish immediately these drops of liquid fire, the snapping and sputtering of the burning watchspring, the showers of sparks and the cracking of glass, all united to furnish a spectacle of startling strangeness. Holding his hands before his face to protect it, the boy was evidently expecting some terrific explosion. But all ended very quietly, and only the bottle, cracked in several places, suffered any damage from this chemical celebration. Then Uncle Paul broke the silence that had ensued.

"Well, Emile, does iron burn? Are you convinced at last?"

"I shall have to be," he replied. "Iron burns, and it burns fast. It was like a little show of fireworks."

"And you, Jules,—what do you think of my experiment?"

"I think it even finer than the one with magnesium. That metal made a light such as I had never seen before; but magnesium itself was something new to us, and so it could not surprise us very much to see it burn. With iron the case is different: we are used to this metal and have so often seen it resist fire that when we see it burn like wood-shavings it strikes us both as something wonderful. But what surprised me most was to see those drops of melted iron stay red-hot for some moments under water."

"Those globules that fell from the spiral as the flame ascended are not iron, but oxid of iron, formed by the combination of that metal with oxygen. I will take out of the bottle those that are not stuck to the glass. They consist, you see, of a black substance that crumbles easily in the fingers. If they were of iron alone, they would not do this. Their softness points to the presence of another element, and this element, as I said, is oxygen. You will find this same oxid of iron in the tiny scales, black and easily broken, that fly off when the blacksmith hammers red-hot iron on his anvil. Both are iron that has been through fire, iron that has become oxidized. Notice also, on the inside surface of the bottle, a light layer of fine reddish dust that was not there before. What can this red dust be? What does it look like?"

"It looks a good deal like iron-rust," replied Jules; "at least it has exactly the color of iron-rust."

"And it is iron-rust,—that and nothing else. Remember this little fact, for it will be useful to you later: iron-rust is iron combined with oxygen."

"Then were there two oxids of iron made in the bottle?"

"Yes, two, but in very unequal parts. The more abundant is the black substance; the other is the red dust deposited on the inner surface of the glass; and this latter is richer in oxygen than the other. I will not dwell on this subject further now, as I shall come back to it later. Notice, finally, the cracks in the bottom of the bottle and the globules of oxid embedded in the thick glass."

"Those drops of oxid must have been terribly hot," said Emile, "to melt the glass like that after going through water. I 've often seen drops of fire fall from the fat when you singe a roast, but I never knew before that there could be ever so much hotter ones."

"Was I right, then, to leave some water in the bottle?"

"I should say so! If you hadn't, the bottom would have been bored clear through."

"More than that; the bottle would have been shattered to pieces by the sudden intense heat. The first drop to fall from the ribbon would have ended the experiment by breaking the bottle. But with this protecting layer of water our bottle has held together and, although cracked, can still be used."

There still remained a fourth bottle of oxygen. Also, safely caged and well supplied with bread crumbs, the sparrow was watching the proceedings. In the midst of plenty, captivity did not seem to depress its spirits unduly. But now its turn had come to be experimented upon, though without any fatal ending in prospect this time, Uncle Paul had assured his hearers. The death of the bird's unlucky predecessor had shown the boys that nitrogen is unbreathable, and that in this gas in which a flame goes out life also is extinguished. What new truth was this sparrow to teach them? It was about to show them the effect of oxygen when breathed unmixed with any other gas. Their uncle took the sparrow and put it into the remaining bottle of oxygen.

At first nothing unusual occurred. Then, after a short interval, the saucy bird became even more alert, brisker in its movements, livelier in every way, than under natural conditions. Hopping about, flapping its wings, stamping with its feet, pecking the glass walls of its prison with furious beak, the little creature was evidently in a burning fever which was fast using up its strength. It panted as if its little breast would burst with the wild pulsations of its heart. Its open beak denoted extreme fatigue, and yet the feverish restlessness still increased. To prevent a sad ending to this scene, Uncle Paul hastened to put the bird back into its cage, where the fever subsided in a few minutes.

"My demonstration is finished," he announced: "oxygen is a breathable gas; an animal can live in it; which is not the case with nitrogen. But life goes on more intensely in oxygen than is altogether agreeable, as we have just seen from the sparrow's extraordinary agitation."

"Never before," said Jules, "have I seen a sparrow so worked up. It acted like one possessed. Why did you take it out of the bottle so soon?"

"Because it would have killed the bird to keep it in there much longer."

"Is oxygen a gas that kills?"

"On the contrary, it gives life."

"Well, then, I don't see—"

"Recall the lighted candle that was lowered into oxygen. It went on burning there, but with a devouring ardor and an immoderate expenditure of wax. The flame was of superb brilliance and vigor, but of short duration. The fuel that would have kept it going a long time under ordinary conditions was used up in a few seconds. It is much the same with life; it goes at an unnatural pace in pure oxygen, uses itself up too rapidly to last long. We might express it by saying that the animal machine is geared too high, and hence, like all over-driven machines, breaks down and stops. You saw how the bird performed all sorts of mad antics, as if violently intoxicated. At that rate its poor little machine would surely have gone to pieces very soon, and that is why I took the exhausted creature out of the bottle, wishing to keep it for another and final experiment. Take good care of it until to-morrow."

## CHAPTER XV

### AIR AND COMBUSTION

THE next day the sparrow, quite recovered from its trying ordeal of the day before, did credit to Emile's careful ministration by its vigor and appetite. The supply of oxygen being used up, Uncle Paul had his nephews prepare, under his supervision, a little more of that gas and also some nitrogen. Burning phosphorus under the bell-glass gave nitrogen, and the decomposition of chlorate of potash gave oxygen. The boys fairly beamed with joy at being allowed to take a hand in these momentous operations. All went off according to rule and with great success. It is true that their uncle was there all the time, advising and directing; but it is also true that both Jules and Emile are unusually skilful with their hands. And so their uncle was not afraid to trust them with bell-glass, balloon, tubes, and bottle, for in hands so careful there was no danger of breakage. When the two gases had been collected the lesson began.

"Oxygen is the only breathable gas," said Uncle Paul, "the only one that will sustain animal life, and also the only one that will make fire burn. But its energies are too powerful, as was proved to you yesterday by the sparrow and the candle. These energies must be toned down by the addition of an inactive gas. When a wine is too heady, we weaken it with water to make a drink that will not injure our health. In the same way oxygen, too strong for breathing or for ordinary combustion when pure, must be weakened with nitrogen, an inactive gas. This mixture gives us atmospheric air, in which nitrogen plays the part corresponding to that of water in diluted wine.

"Our burning of phosphorus under the bell-glass showed us that air is composed of two elements, oxygen and nitrogen, the quantity of the latter being four times that of the former. Now we are going to reverse the operation and make air out of the two elements here before us. Here in this bottle is oxygen, and there in that other is nitrogen. By mixing these two gases in the right proportion we ought to get air like that in which we live, air in which a candle will burn calmly and an animal breathe without danger. How are we to proceed in order to obtain this result? Our unduly strong wine,—that is, our oxygen,—must be diluted with a good deal of water, or nitrogen. In fact, we must add to the oxygen four times its volume of nitrogen.

"Nothing could be simpler. I fill the bell-glass with water, and then cause a bottleful of oxygen to displace a part of the water. The bottle which is to serve as a common measure for both gases I select at random, taking care, however, that it shall be of moderate size, so that the capacity of the bell-glass may not be exceeded when the two gases are mixed. Here we have, then, our oxygen in the bell-glass; and now I fill the same bottle with nitrogen, which I release inside the bell-glass, and I do this four times. This accomplished, the bell-glass contains five bottlefuls of gas,—four of nitrogen and one of oxygen, such being the proportions taught us by our experiment with phosphorus. Consequently, we have here a volume of gas in no way different from the air we breathe, as will be proved in the clearest possible manner by the two experiments we are now about to perform.

"With the gaseous mixture in the bell-glass I fill a gage or a small bottle, into which I lower a lighted candle. The candle continues to burn with its usual calmness, neither faster nor slower than in ordinary air. It behaves inside the gage just the same as it did outside. Our too strong wine is diluted just right. Diluted with nitrogen, the devouring oxygen shows by no means so keen an appetite; it consumes the candle quietly instead of making a little bonfire of it.

"Let the sparrow tell us the rest. I transfer the gas in the bell-glass to a large bottle with a wide mouth, after which I put in the bird. Does anything unusual follow? Nothing, you see. The little captive, transferred to a new prison, is rather disturbed and tries to escape, but shows no sign of painful breathing. Its breast rises and falls as usual, its beak is not open in sign of panting, there is no agitation to indicate suffering; in short, the sparrow breathes in its glass cage exactly as it did in its wicker one, thus proving that the air within is of the same kind as that outside. But to make it still plainer to you that in this artificial atmosphere, the product of our skill, there is no danger of death, we will let the bird stay in the bottle a few minutes longer."

This, accordingly, was done. The boys, in some anxiety as to the result, watched the sparrow closely and were surprised to see its liveliness continue undiminished in an atmosphere made by themselves. The calmness shown by their uncle, who would have been the first to put an end to the experiment if there had been any risk to the patient, reassured them although they still had some slight misgivings, so deeply had they been impressed by the painful end of the other sparrow in the bottle of nitrogen.

"That will do," said their uncle at last. "We know all we wished to know. Set the captive free."

Jules held the bottle, open, out of the window, and the bird flew away as if nothing unusual had happened to it. With a few strokes of its wings it was on a neighboring roof, perhaps giving its comrades an account of the strange things that had taken place in the chemical laboratory.

"What is it saying to them?" Emile wondered. "Is it telling them about its glass cage and its crazy behavior and high fever in the oxygen?" Then to his uncle: "So the air the sparrow came out of is just the same as what we breathe?"

"Yes, just the same. Composed, like it, of oxygen and nitrogen in the proportions I have already named, it maintains the candle-flame and also the life of the breathing animal. With oxygen and nitrogen we made air exactly like the air that keeps us alive."

"Then the air the sparrow breathed we could breathe too?"

"We could breathe it without perceiving the slightest difference, for, as I tell you, it is the same thing."

"I asked because I thought it so strange we could live in air made by our own hands, with our drugs and our outfit of bottles and tubes. And there's something else, stranger still, that came into my head. Let me tell you what it was. Our oxygen here was furnished by a salt, chlorate of potash, in which chemical combination had stored it up. You have told us that there are many other salts, all rich in oxygen, from which this gas could be obtained if it were n't so hard to decompose them. One of these seems to me particularly interesting, the one used for building houses."

"You mean limestone, carbonate of lime?"

"Yes, carbonate of lime. That salt has oxygen just like the others, has n't it?"

"Without a doubt. What of it?"

"Well, if limestone has oxygen, the oxygen could be taken from it?"

"If absolutely necessary the thing could be done, but I warn you it would be enormously difficult in practice."

"No matter; it could still be done. Then we may think of chemistry as telling us that limestone can be breathed, and the idea of limestone as just so much air that we could breathe strikes me as rather funny.''

''You go too far in imagining limestone could be made to furnish air such as we breathe; but it could undoubtedly furnish oxygen. There is no impossibility in that.''

''Could we really breathe air that is made partly out of limestone?'' asked Jules, as surprised at his uncle's answer as at Emile's queer idea.

''Why not? The sparrow, with more delicate organs of breathing than we, breathed air containing oxygen from chlorate of potash, another mineral substance,—in fact, another kind of stone. To accustom you a little to these curious changes and shiftings in which certain elements are used for one thing to-day, for another to-morrow, and for still a third the next day, with no ultimate loss or gain of a single particle of matter, listen to what I am going to tell you now that the opportunity occurs.

''When the lime-burner fires the limestone in his furnace, the carbonic acid which it contains, and which is an invisible gas, escapes and is scattered far and wide in the atmosphere. Vegetables and plants and trees feed on carbonic acid through their leaves. I simply state the fact here, postponing the demonstration until later. They take in from the air the carbonic acid coming from a thousand sources, the least of which, hardly counting at all, is the lime-kiln. They break it up, keep the carbon, and reject the oxygen in a pure state. This oxygen spreads through the atmosphere, becoming a part of our breathable air. Who, then, would venture to deny that in a whiff of air breathed by us there may sometimes be a little oxygen from limestone,—from building-stone, in fact? The gas from stone such as is used in building may, indeed, sometimes help to keep us alive. The elements come and go between one compound and another; substances cease to be, and in doing so give up their material to new substances; the indestructible elements, released from one combination, reappear with properties unchanged in another. Whether it comes from air, chlorate of potash, plaster of Paris, iron-rust, marble, or limestone, oxygen is always oxygen, nothing more and nothing less, provided it be disengaged from all chemical union. Thus the same gas can in turn rust a piece of iron, reduce a stick of wood to ashes, feed a flame, incorporate itself in a wayside pebble and there lose its activity, or send the blood coursing through an animal's veins. Who knows whence comes the carbon in a mouthful of bread? What part may it not have played before entering into the wheat, and what part may it not play afterward? We get lost when we try to follow in imagination the travels of a bubble of oxygen or a lump of coal through all the things that are being continually made and unmade.

"We will not dwell further at present on these wonders, but will return to our subject of air artificially obtained. When, just now, I put oxygen and nitrogen into the same bell-glass, did anything remarkable take place? No; there was not the least rise in temperature, no light, no tumultuous conflict between the two elements,—nothing, in short, that ordinarily accompanies chemical union. Brought together, oxygen and nitrogen did not act on each other; hence, they did not combine chemically in the resulting atmospheric air, but merely mixed. Now, I assure you there is an immense difference between the chemical combination of two gases and their simple mixture. There is an extremely powerful liquid, an acid, which eats into and dissolves most metals, even the hardest, with as much ease as water dissolves sugar; and it is appropriately called *aqua fortis* (the Latin for *strong water*). Its chemical name is *nitric acid.* The Latin term is very expressive, for very few substances can resist the furious strength of this liquid. Our skin, touched in any spot by a drop of it, quickly turns yellow, dies, and peels off in shreds. That is what results from the chemical combination of oxygen and nitrogen. By merely mixing the two gases we get atmospheric air, on the uninterrupted supply of which to the lungs depends our very life; but the chemical union of these same gases produces something that kills. I ask you to note especially here the utter difference between two substances that are nevertheless both made of the same elements. This difference is not unlike that observed by you between the simple mixture of sulphur and iron filings and the chemical combination of these two substances, the combination having none of the properties of either sulphur or iron.

"Thus, air is a simple mixture of oxygen and nitrogen in the proportion of four liters of the latter to one of the former. Oxygen maintains combustion and respiration; or, in simpler language, it makes things burn and animals breathe; but nitrogen merely moderates the powerful energies of the oxygen mixed with it in the air around us. What takes place in the act of breathing deserves our serious study, but the proper time has not yet come for that. Some day, after we have learned certain things that will prepare the way, we will take it up again, in detail. At present let us confine our attention to the subject of combustion, or burning, and especially to that seen every day in our own fireplaces. A substance burns when it combines with oxygen, and so in every act of burning there must be both something to burn and oxygen to make it burn. Let us look into this a little more closely.

"When we wish to make a fire burn more briskly, what do we do? We take the bellows and blow air on the fuel,—the wood, coal, or charcoal. At each blast from the bellows the fire revives and its strength increases. The live coals, at first of a dull red, grow bright red and then glowing white. Air brings new life to the fire by giving it oxygen. But if we wish to keep the fuel from burning up too fast, what do we do? We cover the fire with ashes and thus guard it from too free contact with the air. Under this covering the coals remain alive for a long time, being only very gradually consumed. Thus fire is kept up in a fireplace only by continual supply of air, the oxygen of which combines with the fuel as combustion goes on.

"If a fire is to be brisk and give out good heat, air must be supplied in a rapid current proportional to the amount of the fuel. In the economical foot-warmer, heated by a few live coals, there is but little air admitted, and it gets at the coals only through a covering of ashes. Combustion is proportionately slow, and the heat given out is slight, but it is lasting. On the other hand, in the great blast furnaces of our iron-works, consuming their fuel by the cart-load, air is supplied in powerful gusts by blowing-machines that raise a veritable tornado. This hurricane of air fans something more than a brazier of live coals; it creates a sort of roaring inferno. Call to mind our sitting-room stove, when, having been first cleaned out and then well filled with fuel, it burns with a kind of subdued rumbling."

"I know what you mean," Emile broke in; "we say then that the stove is snoring."

"Yes, and it is the cause of this snoring that I now wish to explain to you. If the door of the ash-pit is open, or at least partly open, the stove snores; but if it is closed the stove is quiet. Why is this? Evidently because something rushes noisily into the stove when an entrance is left for it. What this something is will not be hard to find out. Hold your hand near the door of the ash-pit and you will feel a lively current of air. So it must be air that goes with a snoring sound through the bed of burning coal. That is what we call a draft. A stove that snores has a good draft; that is, plenty of air passes through its burning fuel, and so the fire is vigorous and gives out a good heat. A silent stove has a poor draft; air comes in but slowly, and the fire is low. According to the freedom or hindrance with which air is admitted,—that is, according to the strength or weakness of the draft,—the fire burns rapidly or slowly.

"Now let us seek the cause of this draft. Over a hot stove, wave a piece of burning paper, and you will see the burnt particles rise in an eddy, going to a greater or less height, sometimes even up to the ceiling. Those bits of burnt paper, light though they are, do not go up like that of their own accord; an upward draft must carry them. This draft is produced by the ascending flow of air that has been warmed by contact with the stove and thus made lighter; it rises and is immediately replaced by cold air, which in its turn becomes warm and rises. Although air is invisible, its ascent can be inferred from the rising particles of burnt paper that are carried up with it, very much as the imperceptible movement of nearly quiet water is shown by the drifting of objects floating on its surface.

"There is another experiment I advise you to perform next winter, when the stove is going. Take a sheet of paper and cut out a circle as large as your hand; then with a pair of scissors cut this circle into a spiral ribbon, following a line that starts from the edge and gradually approaches the center. Attach the middle of this spiral to the lower end of a wire hung vertically over the stove, and then let go of your paper ribbon. It will stretch out, by its own weight, into a sort of corkscrew, big at the bottom and small at the top where the wire supports it. If the stove is hot you will see the corkscrew whirl round and round like some ingenious piece of mechanism. The cause of the whirling is this: The paper ribbon presents its surface diagonally to the current of hot air that is continually rising, and from the push it thus receives all along its length comes the operation of the little mechanism. It is this same push of moving air against the diagonal sails of a windmill that causes them to turn.

"Thus it is proved that on being heated, air becomes lighter and consequenty rises, while cold air rushes in to take its place. The push of this rising air turns our paper corkscrew, and it is this same current that carries upward the bits of burnt paper. You will now be able to understand what takes place in our stoves and fireplaces when we say that they are drawing well. If the air were all of the same temperature in the chimney, in the room, and out of doors, there would be no draft. But as soon as the fire is lighted, a change of conditions is brought about: the column of air in the stovepipe or in the flue of the chimney gets warm, becomes lighter, and rises. The hotter the air and the taller the column, the faster the ascent. As the hot air rises, cold air, which is heavier, rushes toward the fire, makes it burn, becomes heated in doing so, and passes up in the ascending column. In this way a continual current of air is established from the bottom of the chimney to its top. In passing over and through the burning fuel, this constantly renewed air-current feeds the fire with its oxygen, and as soon as it is heated and has taken on its load of carbon continues its journey up the chimney, taking the smoke along with it and finally escaping into the open air. That is how a chimney draws, how a stove snores. The draft acts like a pair of bellows working of its own accord, renewing the air as fast as the oxygen is used up, and so keeping the fire going. To maintain a good fire, follow this simple rule: allow free entrance of air with its fresh supply of oxygen, and free exit of air that has been used and no longer has any oxygen, or at any rate has not enough of it. Let plenty of air come in below, let it circulate without hindrance through the burning fuel, and then let it pass upward and give place to a fresh supply. By so doing you will have a fire that, with sufficient fuel, will do you credit."

## CHAPTER XVI

### RUST

THE boys had just found an old rusty knife-blade in the garden. A few weeks earlier they would not have paid the slightest attention to this useless scrap of iron: it was not worth looking at, much less picking up; but since their uncle had told them about the combustion of metals they were looking at things in a different light, and so this scrap of rusty iron was considered worthy of examination. There is nothing like knowledge to give food for thought. What the ignoramus disdains to look at, the well-informed person picks up, examines, and often finds deserving of serious study. So the old knife-blade was picked up by Jules, who at once noticed the close resemblance between the reddish iron-rust and the fine powder with which the burning of iron in oxygen had coated the inside of the bottle. He called his brother's attention to this resemblance.

"Here is a scrap of iron," they said to each other, "that no one ever thought of burning in a bottle full of oxygen, and yet it has turned to rust just as the watchspring did after being set fire to with a bit of tinder. How did it happen? Let 's go and ask Uncle Paul."

In reply to the question, their uncle said to them at lesson-time:

"Most metals, if left to themselves after being polished, will gradually become tarnished, taking on a sort of coating very different from the original luster. If you cut a piece of lead with a knife, the section shows bright and shiny; but this brightness soon fades, and the cut tarnishes until after a while it looks as dull as the rest of the lead. Something of an ashy-gray color has overspread the shining surface. So with iron or steel: how it shines when it comes freshly polished from the manufacturer's hands, and how little of that shine is left after it has been for some time exposed to the air! At first of a brilliant luster almost rivaling that of silver, it gradually becomes covered with reddish spots that grow in size every day, until finally they spread over the whole surface and eat into the metal. This process we call rusting, and sooner or later the metal will be completely changed to a red earthy substance. That is what happened to the knife-blade you found in the garden.

"Lead also rusts, in its own way. Instead of turning into a red earthy substance, it turns into a grayish earthy substance. The dull coating that so quickly tarnishes the fresh cut in lead is the beginning of rust. In time this coating would become a thick earthy layer. Zinc rusts in like manner: on the surface it is of a dull gray; within, of a shiny white. Copper keeps its polish no better, becoming overlaid after a while with a green coating due to the rusting of the metal. So it is that our commonest metals all have a serious defect: they lose the brightness so pleasing to the eye and turn into a crumbling earthy substance. In a word, they rust.

"Such is the fact. What is the cause? We need not look far to find it. You have seen iron burning in oxygen and powdering the inside of the bottle with a fine reddish dust having all the appearance of rust. In fact it is rust, nothing else. When we put a few pieces of zinc into an iron spoon, the other day, and thrust the spoon into a bed of glowing coals, you saw the zinc melt, catch fire, and turn into a white flaky substance. That substance is zinc-rust. Lead, if kept melted long enough in a furnace supplied with a current of air, changes into a yellowish earthy substance, which is lead-rust. A sheet of copper held in the fire loses its red color and turns black, at the same time giving the flame a beautiful green tinge. The black material thus formed is copper-rust. In short, these different rusts are all burnt metal: they come from the combination of these various metals with oxygen; in other words, they are oxids.

"So far we have been on familiar ground, thanks to our experiments in the burning of metals; but here we come to something new. These oxids that come into being amid such dazzling splendors of light and such generation of heat—these rusts that have their birth amid gorgeous displays of fireworks—are in no wise different from those common rusts that slowly tarnish the surface of metals. When a piece of iron, buried in the damp earth, slowly becomes encrusted with reddish matter, and when another piece of iron burns with a brilliant light in a bottle filled with oxygen, it is the same chemical action going on in both instances. When one bit of zinc takes on a grayish coating and another, melted in an iron spoon, burns with a beautifully colored flame, leaving white flakes as the result of the combustion, the process is essentially the same in the two cases. In both, the oxygen of the air combines with the metal. Common rust is an oxid, a burnt metal, and whenever and wherever it forms there is real combustion, whether heat be sensibly present or not. A few further examples will not be out of place here.

"Exposed to the air for a long time, a piece of wood is gradually consumed: it turns dark, and finally crumbles to a brown dust. This crumbling of the wood is really slow combustion, differing only in its gradual accomplishment from the combustion attended by fire. The rotting wood combines with the oxygen of the air as burning wood does in a fire-place, and, like burning wood, it even gives out heat. We all know something about the heat that is thus produced. The inside of a dung-heap is decidedly warm, and a damp haystack may become so heated as to take fire. In both cases there is a slow burning of vegetable matter under the action of the oxygen in the air. So, too, with decaying wood: it is in slow combustion, and it gives out heat.

"Why this heat is not felt can be easily explained. Suppose a log takes ten years to be consumed by the slow process of decay, and that a similar log would burn to ashes in an hour. In both cases heat would be generated, but with the rotting wood this heat would be given out very slowly, as it must spread itself over ten years. Naturally, then, it would be too slight to be felt. With the log that is burned in the fireplace, however, heat would be rapid and abundant, for it would all be crowded into one hour of time, and could not fail to make itself felt. It is clear therefore that, although at bottom the chemical action is always the same, there are many different degrees of rapidity with which things may burn. An old decaying tree trunk, a heap of steaming dung, a damp haystack smoldering within, a stick of wood blazing on the hearth—all these are so many examples of slow or rapid combustion; in them all, the oxygen of the air combines with solid matter that can burn, and the only point of difference lies in the rate of combustion. There is rapid and there is slow combustion,—rapid when the substance is ablaze and burns with much heat and light, slow when it is consumed gradually and without any flame or light, and often with no perceptible heat. The first kind of combustion is brilliant but short; the second makes no show and is much longer.

"Rust, then, is to metals what decay is to a heap of vegetable matter: it is the result of slow combustion. Exposed to the air, especially to damp air, metal combines with oxygen, becomes oxidized, as we say; that is, it turns into the compound we call an oxid. This explains why the old knife-blade took on a reddish crust, why freshly cut lead tarnishes almost immediately, why zinc, lustrous inside, wears a grayish coating on the outside. That red crust is oxid of iron, that dull film on lead is oxid of lead, and that grayish coating is oxid of zinc. A longer or shorter contact with moist air has sufficed to burn the metal, at least on the outside.

"Nearly all metals act thus. Eaten by the oxygen of the air, they turn to rust, yellow or red for iron, green for copper, grayish white for lead and zinc. Not all turn with the same readiness. Among common metals iron rusts the quickest; then come zinc and lead; next in order are copper and tin, and after them we have silver, which can be kept untarnished a long time. Gold is an exception: it never rusts, and it is precisely this quality of always retaining its luster that makes it so valuable. Gold coins and ornaments have come down to us from the earliest times as clean and bright as if made only yesterday, despite the long centuries they have lain in a damp soil that would have reduced other metals to nothing but rust."

# CHAPTER XVII

## AT THE BLACKSMITH'S

ONE day Uncle Paul took his two pupils to the village blacksmith, whose smoke-stained shop was to serve as laboratory for one of the most curious experiments of their course in chemistry. He wished to prove to them that water contains a highly inflammable substance, a gas easier to burn than even phosphorus and other elements they had seen catch fire so readily. Water, which puts fire out, was now to furnish fuel for fire. Jules and Emile did not show themselves very confident of the success of what they could not but look upon as a rather foolish undertaking. The blacksmith himself —farrier for half the week, locksmith on occasion, horse-doctor now and then, maker of cutlery if necessary, plumber in spare moments, tin-plater when old saucepans required his services, silversmith and even jeweler at a pinch (for one must make a living, said he, somehow or other)—the blacksmith, we say, when told of the affair in hand, found nothing in all his varied experience to encourage hope of success in his neighbor's proposed enterprise. Nevertheless he lent a hand in the necessary preparations with a very good grace, and placed his forge, tools, and personal assistance entirely at the other's disposal. The coal-dust that begrimed his face partly hid his mischievous smile of incredulity.

A large earthen bowl filled with water was placed on the work-bench, together with a tumbler, and a heavy iron bar was thrust into the forge to be heated red-hot. The blacksmith plied the bellows, and Uncle Paul watched the iron bar. When it was hot enough, he directed the others how to proceed.

"Fill the tumbler with water," said he to Jules, "and then with one hand hold it upside down in the bowl, the mouth always immersed. I will plunge the red-hot end of the iron bar into the water and under the inverted tumbler. Don't be afraid for your fingers; I will take care not to burn them. Without raising its mouth out of the water, tilt the glass so that the red-hot iron can be slipped under."

All this being made clear, Uncle Paul quickly plunged the end of the bar, heated to the utmost, under the mouth of the tumbler. The water boiled and bubbled violently for a moment, while globules of gas were seen to rise and collect in the inverted bottom of the glass.

"That is not enough," said Uncle Paul. "Keep hold of the tumbler while I do it three times more, always the same way, until we have several fingers' depth of gas in the glass."

Again and again the bar was returned to the forge and then plunged, glowing hot, into the water; and with each repetition the volume of gas increased. It went slowly, but still it went, and the blacksmith worked the bellows untiringly, being as eager as the boys to see the result of this curious experiment. What could there be collecting in the glass? It seemed to be a sort of air, and of course it was invisible. But how did it differ from the air outside? In the course of his daily work the smith had often put red-hot iron into water, and the hissing sound that followed was familiar enough to him; but he had never gone any farther. It was only those that read books—neighbor Paul, for instance—that thought of collecting in a glass bubbles from water boiling at the touch of red-hot iron. The soot-begrimed face, streaming with sweat that looked like drops of ink, had by this time lost its incredulous smile, and in its place was an unmistakable expression of deep interest. The forge that had seemed to the worthy man to hide no secrets from him, did, after all, hold mysteries that would no doubt soon be explained.

At last Uncle Paul himself took the glass in one hand, tilted it slightly so as to let the gas escape little by little, and with the other hand held a lighted paper to the bubbles as fast as they appeared on the surface of the water. Immediately there was a pop from the bursting bubble, and a flame darted up, but so pale that one had to turn one's back to the door in order to see it. The dark shop was, in itself, well suited to the purpose of making the burning gas apparent to the eye. *Pop!* went the second bubble, and *pop! pop! pop!* repeated the others in quick succession, each giving out a feeble flash of light. It was a sort of miniature fusillade.

"Waterproof gunpowder!" exclaimed the blacksmith, in astonishment. "No sooner does it come to the surface than it explodes. Once more, please, so that I can see it better."

Uncle Paul again tilted the glass. *Pop! pop!* went the bubbles until the gas was all used up.

"And you say," inquired the farrier, "that this air, this gas that catches fire quicker than gunpowder, comes from water?"

"It comes from water decomposed by the red-hot iron. What else could it come from? In getting it I use only water and iron, and even the latter is not really necessary, as we shall presently see. It is, then, the water that gives us the inflammable gas."

"What a fine thing chemistry is!" said the smith, with a puzzled shake of the head. "It makes water burn. I 'd like to study chemistry a little if I had time."

"You practise chemistry every day," rejoined Uncle Paul, "and very interesting chemistry, too."

"Chemistry—I? Is it chemistry when I shoe Jacques's mule or sharpen Simon's plowshare?"

"Yes, there is chemistry in those things, and you are putting chemistry into practice every day, but without knowing it."

"Well, that beats me!"

"I hope to make it all plain to you before long."

"When?"

"Now, this very day."

"One question more, if you please, my learned neighbor. What do you call this gas that comes from water and burns?"

"It is called hydrogen."

"Hydrogen. All right; I'll remember that word. Some Sunday after vespers I'd like to show a few of my friends what you have just shown me. But go on. A poor ignoramus like me ought not to interrupt by asking questions when you want to be teaching your nephews. They are lucky little chaps to have you for their teacher. Oh, if I were only of their age and you would take me for a pupil! But it's too late, much too late. My old brain could n't make anything of books. Now, what else can I and my forge do for you?"

"Start up the fire again, my good friend, and make a solid bed of red-hot coals, just as hot as you can, with no flame. I am going to decompose some more water, this time with live coals instead of red-hot iron. We shall get the same inflammable gas,—a proof that it is the water that gives it, and not the iron or coal used in the process. You, Jules, hold the tumbler ready. The experiment will be just the same as with the heated iron."

They waited a few minutes to let the forge get as hot as possible; then, taking up a glowing coal with the tongs, Uncle Paul plunged it into the water and under the mouth of the glass. A bubbling followed, the gaseous globules rising in even greater number than with the iron. With a few repetitions of this procedure the glass was nearly full. The collected gas was found to burst into flame at the touch of a lighted piece of paper, though the flame, as before, was very pale; and at each outburst of flame a slight explosion was heard. In short, the live coals served exactly the same purpose as the red-hot iron, showing plainly enough that the inflammable gas, the hydrogen, as Uncle Paul had called it, came from the water; red-hot iron and live coals, things very unlike each other, served merely to set it free by decomposing the water, a little at a time.

The blacksmith seemed lost in thought over what he had just witnessed. He was reminded of an every-day occurrence in his work at the forge. His neighbor did not fail to perceive this.

"Tell me," said he to the smith; "when you want to heat a piece of iron just as hot as you can get it, so as to do a bit of welding, let us say, how do you go about it?"

"How do I go about it? I was this minute thinking of that connection with your hydrogen there, for it seemed to me to explain something that I do every day without understanding the reason. Over there in the corner is a small trough full of water, and in it I keep a sort of little rag mop with a long handle. I use it to sprinkle water on the coal in the forge, and so I get a heat such as I can't get any other way."

"You throw a little water, then, on your fire to make it hotter; you get it to burn faster by using what would seem more likely to put it out."

"That 's what I do, and I 'm always puzzled when I stop to think about it. Now, with your hydrogen, it might be—"

"One moment, please. We will come back to that presently. I see from my nephews' astonished looks that wetting coal in order to make it burn better rather upsets their ideas. Suppose you show them how it 's done.''

''All right. I 'm glad to do anything I can for you, to balance what I 'm getting out of this lesson, now that I 'm lucky enough to be one of your pupils for the day.''

Plying his bellows once more, the blacksmith started up the fire, on which he had already heaped fresh coal. An iron bar was thrust into the glowing mass, and after it had been for some time subjected to the most intense heat possible under these conditions, it was withdrawn by the smith.

''See,'' said he, ''It 's red-hot, and I could n't make it any hotter by just leaving it there and pumping the bellows. It 's as hot as I need it usually. But if I want to get my iron hotter still, so as to weld two pieces together on the anvil, I take my sprinkler and throw a few drops of water on to the fire; but not too much, you understand, for that would put it out.''

The iron bar was then put back into the forge, and the coal was slightly sprinkled with water. The boys, like two apprentices, stood one on each side of the blacksmith, intent on seeing everything that was being done. A trivial operation that they must have witnessed many times, but without paying it the slightest attention, took on for them an intense interest now that their uncle had opened their eyes to the properties of hydrogen, the inflammable gas contained in water. To acquire an interest in anything there is nothing like having your attention called to it. Knowledge adds charm to everything around us.

The effect of the water on the live coals was immediate. Behold, the tongues of flame, at first full and long, and of greatest brightness at the bottom, and reddish and smoky at the top, suddenly shrank and seemed to withdraw into the midst of the burning fuel. Then here and there, through the openings in the mass of coal, sprang up short jets of flame, giving a clear white light. These little tongues of white flame were not unlike the hydrogen flames that had been so hard to see in the daylight. Their temperature was evidently very high, for the glowing mass whence they sprang dazzled the eye. Again the bar of iron was withdrawn, this time not red-hot but of a blinding white. Snapping and crackling, it sent out a splendid shower of sparks.

"It 's just as it was when we used oxygen," declared Emile, drawing back to avoid this outburst from the metal fireworks: "the iron burns."

"Yes, my little friend," replied the blacksmith, "the iron burns, and it burns so well that if I left it too long in the forge with the heat as it is now, you would see my iron bar get smaller and smaller till there was n't much left of it. Look round on the floor near the anvil, and you 'll find plenty of little bits of burnt iron. We call them splinters; the hammer strikes them off from the hot iron."

"I know what you mean: they are oxid of iron."

"I don't know that word, but those little splinters are iron that has burned. They are plenty enough when I make the forge as hot as I can by sprinkling on water. But let 's hear now what your uncle has to tell us. How is it, neighbor, that water can start up a fire like that? Without water the iron only gets red-hot in the forge; with water it turns blinding white. That 's what I don't understand."

"You will understand," replied Uncle Paul, "when I tell you that hydrogen is the fuel that gives the most heat. Neither wood, coal, charcoal, nor any other fuel makes so hot a fire as hydrogen. It is the very best of fuels, having no equal for readiness to take fire and for intensity of heat produced."

"Now I understand; at any rate I think I do," rejoined the blacksmith. "I throw a little water on the burning coal in my forge, and the water is decomposed, as you call it, just as when I saw you plunge a red-hot coal into the water under the glass. Hydrogen is produced, and it mixes with the coal and burns; and as it is the best of fuels it makes the intense heat that turns the iron white-hot so that it can be welded. With my sprinkler I give the fire fuel that is better than coal. Is that the way of it?"

"Precisely. Water, decomposed by the live coals, furnishes the fire with additional fuel, the very best. Did n't I tell you that you were practising chemistry every day, and very scientific chemistry, too?"

"My word, you 're right! But I never suspected it. How was I to know that by wetting my coal I was making hydrogen? One has to read books to know those things, neighbor; but I am ignorant and have to give my time to my hammer and anvil, not to printed words. One thing more, now we are about it. I 've heard it said by persons of education that when a fire gets well to going, it 's a bad plan to try to put it out with water unless you have plenty of it. If you have n't, it is better to smother the fire,—with earth, for example. Does hydrogen have anything to do with that?"

"Most assuredly it does. If a little water is sprinkled on a hot fire, the water is decomposed, and so furnishes hydrogen as additional fuel for the fire, which, instead of being put out, burns all the faster, just as your forge gives more heat when you throw on a little water. But if, instead of moistening your coal with a gentle sprinkling, you watered it freely, by the bucketful, your fire would be extinguished. So fire must be fought with water that does not have to be measured out by the cupful; otherwise it is like pouring oil on to the flames, as they say."

"One need n't talk with you more than five minutes to learn something new," declared the blacksmith. "My forge is always at your service, as you know. Only too often it stands quite idle, so slack is my trade. If your chemistry calls for any more of my outfit, help yourself. My tools are yours to do what you like with."

Uncle Paul thanked his friend and took his leave. Jules carried away with him, to examine at his leisure, a handful of the iron splinters he had picked up around the anvil.

On their return home the boys asked permission to perform for themselves the fine experiment they had just seen at the blacksmith's. The inflammable gas coming out of water had so astonished them that they wished to see it again, and above all to try making it themselves, with no help from their uncle; and in fact it was one of the simplest operations, nor did it require the use of any dangerous drugs. The blacksmith, it is true, had shown himself most obliging; but they did not like to make too great demands on his time or his good will. Besides, there might be some mule waiting at the door to be shod, or some tool to be mended might be heating in the forge. In busy moments of that sort chemistry would only be a bother to the good-natured smith. Home was by far the best place for it. Without disturbing any one they could perform the hydrogen experiment over and over again at their pleasure. But was the thing feasible?

"Quite feasible," their uncle assured them. "Get a brazier and light some charcoal; that will do as well as coal, and even better. Have a bowl full of water, and a tumbler, and proceed exactly as we did at the blacksmith's. When your coals are red-hot, take them up, one after another, with the tongs, and plunge them quickly into the water under the mouth of the glass. In that way you will obtain inflammable gas just as well as with the glowing coal from the forge. To be sure of the best success, see that your live coals are as hot as you can make them with the bellows; for the hotter they are, the more water they will decompose. Finally, let me caution you to look out and not burn your fingers."

"Oh, no fear of that," said Jules. "Emile will hold the glass and I 'll manage the coals. I sha'n't be such a blunderbuss as to burn my assistant's hand."

"I warn you that if you try to operate with hot iron, your success will be doubtful, as your brazier is hardly big enough to heat to redness an iron bar of any size. But try it if you wish, and, once more, take care not to burn yourselves."

Having given these directions, Uncle Paul left his nephews to their own devices. So well did they arrange their charcoal in the brazier, and so good a draft did their fire get, that the two young chemists soon had a bed of glowing coals at their disposal. The operation went off as well as could have been desired, the hydrogen bubbling up in fine style. Jules, whose sharp eyes nothing escaped, could even detect, when they set the hydrogen on fire, a pale bluish tinge in the flame, which had not been observed in the hydrogen flame when red-hot iron was used at the blacksmith's. Emile, too, when called upon to notice the difference, did not fail to perceive it.

They then undertook the same experiment with hot iron. But they had to content themselves with something no bigger than a curtain rod, and indeed their fire could not have heated to redness a larger bar if they had had it. Consequently, they were obliged to heat and reheat their slender iron rod over and over again, at the expense of much time and patience, before they got even a small quantity of hydrogen by this method. A few little bubbles of hydrogen, which burned with an almost invisible flame, were the total result of exertions that left them both in a profuse perspiration, so many times did they have to repeat the same operation. But, after all, they accomplished enough, for their uncle had warned them not to expect any great success.

## CHAPTER XVIII

### HYDROGEN

THE use of red-hot iron for obtaining hydrogen from water is a slow and tiresome process, requiring many repetitions of the same operation to secure even a small quantity of the gas. With live coals instead of hot iron, speedier results are obtained, but the hydrogen is not pure; it is mixed with other gases derived from the coals, and to these is due the bluish tinge of the flames, a peculiarity detected by Jules. Excellent, for practical reasons, as are these two simple and easy methods when the sole object is to show that water contains an inflammable gas, they must give place to others when it is desired to obtain a considerable quantity of hydrogen in a short time.

"Let us turn now," said Uncle Paul, "from this way of getting hydrogen from water,—I mean by the use of live coals. What we really get is a mixture of various gases that have to be examined separately if we are to avoid confusion. And let us turn, too, from the method in which red-hot iron is used; though this time the hydrogen is pure, there is very little of it. What we are looking for is some simple process that will give us all the hydrogen we desire, and that without furnace, forge, or brazier, which are not always conveniently at hand. I have to tell you here that iron can decompose water without first being heated if it only has the help of a little sulphuric acid. With these two acting together, the hydrogen in water can be set free with all the ease one could ask for. I must also tell you that another common metal, zinc, is still better than iron for decomposing water,—always, of course, with the help of sulphuric acid. We can, then, use whichever of the two metals we chance to have at hand, though zinc is to be preferred. If that is lacking, we will resort to iron filings, which, as they consist of minute particles, readily yield to chemical action when brought into contact with the other substances used in this method.

"Into this tumbler I put water and some pieces of zinc from the old watering-pot that supplied me the other day with material for showing you that this metal will burn. No results are apparent as yet, all remaining quiet in the glass because cold zinc by itself has no effect on water. But I add a little sulphuric acid and stir it in well. Now things will go on unassisted. The water begins to boil violently, sending up countless bubbles of gas that burst on reaching the surface. These bubbles come from the decomposed water; they are hydrogen, precisely the same inflammable gas that we obtained by using red-hot iron in the blacksmith's shop. Watch now. I hold a piece of lighted paper near the surface of the water, and each bubble, as it bursts, catches fire with a slight explosion, burning with a flame so pale as to be visible only in the dark. As the bubbles follow one another thick and fast, there is an almost continuous popping.''

This minature artillery popping away on the surface of the liquid, and these flames dancing on the water, certainly offered a curious spectacle. But there was something else that appeared to have even greater interest for the young spectators: the water had started to boil with no fire of any sort to heat it, and the glass had become so hot as to make one almost afraid to touch it. Uncle Paul anticipated the surprised inquiries prompted by these remarkable developments.

''Look into the glass,'' said he, ''and you will see that the hydrogen bubbles first make their appearance on the zinc, for it is there that chemical action, resulting in the decomposition of the water, takes place. These bubbles of gas make their way up through the liquid and in doing so cause considerable commotion, just as water boiling over a fire is agitated by the bubbles of steam that are being formed. In reality the water in this glass is not in motion as a whole, but is merely stirred by the uprushing bubbles in the same way it would be stirred if you blew air into it through a straw. The boiling of the water is only apparent, only an agitation that deceives the eye.''

''But the glass is awfully hot,'' remarked Emile; ''I can't bear my hand on it.''

''Very true; but the heat is still far below that of boiling water. If you should ask me to prove it, I should only have to take the tongs and lift out the piece of zinc, whereupon the liquid would immediately quiet down, there being no further generation of hydrogen, which caused the commotion."

"All the same, there 's lots of heat there. Where does it come from, with no fire to make it?"

"I see Emile finds it hard to get used to the idea of heat without fire. Did we need any fire to make the mixture of powdered sulphur and iron filings raise the temperature of the bottle to a burning heat? Does the mason use fire when he pours cold water on lime and makes a paste that is too hot for the hand to bear? Without fire, without live coals, without any apparent cause, great heat is produced in both cases, and chemical combination explains it. In our tumbler, here, we have an example of it. Water is being decomposed, but at the same time the opposite process, combination, is going on between the acid and the metal; and this process generates heat. We will come back to this interesting point later, and you will then see that the heating of the liquid in our glass is only what might have been expected, for zinc is really burning, or undergoing slow combustion.

"It is not enough to know how to get hydrogen with zinc and sulphuric acid; you must also provide something to receive and hold the gas. A little difficulty presents itself at the start. We have to do with three substances,—water to furnish the hydrogen, and sulphuric acid and zinc for decomposing the water and so releasing the hydrogen. All the water and all the zinc to be used in the operation may be put into the glass at once, but the sulphuric acid should be added little by little as it is needed.

If poured in copiously and all at once, it would raise an unmanageable commotion. Under these conditions the operation would be too rapid, and the operator would run the risk of being splashed with the boiling liquid. Consequently, the sulphuric acid should be poured in gradually, a fresh supply being added whenever the release of gas slows up. Moreover, these successive additions of acid should be made without opening or in any way disarranging the vessel in which the hydrogen is generated. In this manner we prevent any admission of air, which would mingle with the hydrogen and form a dangerous mixture.

"The vessel commonly used in this operation is a sort of bottle with two necks, one in the usual central position, the other at one side. Into this bottle is put a handful of zinc cut into small pieces, or, better, a small sheet of zinc rolled up so as to pass through the neck. Enough water is then poured in to cover the metal completely. Through one of the necks, no matter which, is passed a glass tube, which is held in place by a tightly fitting cork stopper with a hole in it to receive the tube, and which is bent over and downward on the outside like the one we used in producing oxygen. Finally, through the other neck and into the liquid is passed a straight glass tube, which is held in place in the same manner as its companion. The apparatus is now ready for use, only sulphuric acid having to be added. For this purpose the straight tube is equipped at the top with a small glass funnel, through which the acid required is gradually introduced. As long as the release of gas proceeds satisfactorily, no further attention is necessary; but if it slackens, a fresh dose of acid is poured in. This arrangement is very simple and very ingenious. The straight tube, extending into the water as it does, admits no air to mingle with the hydrogen, a thing to be carefully avoided, as will be shown; but it does allow the introduction of sulphuric acid whenever needed. Furthermore, the hydrogen that is being released cannot get out this way, as the water keeps it back; hence its only issue is through the bent tube, the

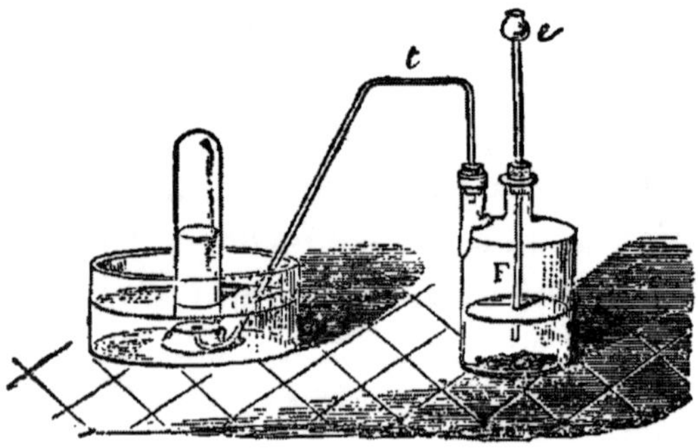


nearer end of which is in the second neck of the bottle and well above the water. In other words, the gas-factory has two doors and only two,—the straight tube, which allows entrance but not exit, and the bent tube, which offers a way out, but no passage inward, when the apparatus is in operation and discharging its hydrogen.

"One thing more. Suppose the bent tube gets stopped up in some way, or that it is too small for a sufficiently rapid discharge of the gas set free by the decomposition of the water; what will happen? The gas collected in the bottle, and unable to get out, will press downward on the liquid and drive it up through the straight tube until it overflows the funnel at the top. This rising of the water in the straight tube warns us that something is wrong with our apparatus, blocking the issue of the gas. But, unless we pour in too much sulphuric acid at a time, we need not trouble ourselves to watch the straight tube for signs of danger.

"Such is the hydrogen apparatus used in laboratories, and I regret that I cannot show you one in operation to supplement my description; but a two-necked bottle is not an easy thing to procure in our village."

"That 's so," chimed in Emile. "I 've never seen anything of the sort around here. A bottle with two mouths, one for pouring in and another for pouring out, is n't a thing you 'd be likely to find in any rubbish heap. And so we can't have our hydrogen factory, after all," he concluded in a plaintive tone.

"But should I have aroused your expectations at the blacksmith's if I had n't known beforehand that I could gratify them? With an old pickle-jar and a little ingenuity, need we despair of success? Your uncle thinks not. What does our modest bottle lack to make it a serviceable piece of apparatus? Two mouths; and we will supply them without further delay."

So saying, he took a good-sized cork stopper that had belonged to some demijohn, and shaped it carefully with a file to make it fit the large neck of the pickle-jar. Then he punched two holes in it, and in one of these he fixed the bent glass tube, pushing it only just through the cork, while through the other hole he passed the straight tube, forcing it down much farther. Bits of zinc and enough water to cover them well were next put into the jar, after which the stopper was inserted, with a little moist clay around the edge to prevent any escape of gas. Emile was delighted with the turn things were taking: he was going to see hydrogen made in as large quantities as any one could wish. Everything was arranged in accordance with his uncle's previous description.

"Here is the straight tube," the boy pointed out in eager interest, "that you 'll pour the acid into when the time comes, and there is the bent one for letting the hydrogen out. The old pickle-jar is going to do very well, now that it has two mouths made in the big stopper. But there 's one thing we have n't got yet,—the little funnel for pouring in the acid."

"I have none," replied his uncle.

"What shall we do, then? The tube is so small we can't pour anything into it without a funnel."

"Let us ask Jules and see whether he would allow such a trifle to defeat our purpose."

"You will laugh at my idea," said Jules, on being thus turned to for advice, "but why could n't we use a little piece of paper rolled up into a cone open at the point?"

"Your suggestion is unanimously adopted. Lacking a regular chemist's funnel, we could hardly do better. Your little paper cone shall take the place of the small glass funnel; but it will soon go to pieces, I warn you, for sulphuric acid is exceedingly destructive. However, that does n't matter in this case, as we can renew our paper funnel as often as necessary. Economy in this particular is not required."

So said, so done. The paper funnel inserted in the upper end of the straight tube made it possible without the slightest difficulty to pour in the sulphuric acid, whereupon the water in the jar immediately began to boil, as it appeared to the eye, and hydrogen came out through the bent tube, the further end of which went down into the water in the bowl. The boys hastened to touch a lighted paper to the gas bubbles thus created. Quick flashes of flame, a crackling sound, a pale white light—all these duly followed as the gas came rushing out of the pickle-jar and was ignited. It was really and truly hydrogen; a regular laboratory outfit could not have given better results.

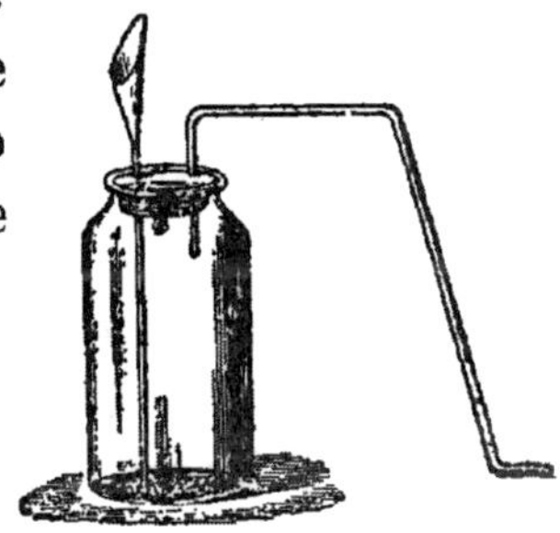

"You are pretty well acquainted with this water artillery now," said Uncle Paul. "Let us pass on to something else and set fire to a large volume of hydrogen. I dissolve a little soap in water, and into this soapy water I lower the end of the tube through which the gas is discharged. If we took a straw and blew through it, we should get plenty of foam. The bottle blows in its own peculiar way:

it sends a jet of gas into the midst of the soap-suds and makes a mass of tiny bubbles, all filled with hydrogen. In this manner we obtain a certain amount of the inflammable gas stored up in little thin-walled cells. I apply a piece of lighted paper, and the gas catches fire. The explosion is louder, the flame larger than before, though the light produced is still very pale."

At the request of the young pupils, who were quite fascinated with this exhibition, the experiment was repeated and a still greater volume of gas was produced, which was then exploded with fine effect.

"We have nothing more to learn from this plaything," concluded Uncle Paul. "It has shown us how readily hydrogen catches fire: hardly do we touch the lighted paper to the bubbles, when the imprisoned gas explodes. Let us now proceed to another experiment, which will show us that hydrogen, so highly inflammable in itself, can yet be used for putting out fire. It burns as nothing else will, and yet it stops the burning of anything plunged, all afire, into the midst of this gas. It will put out a lighted candle as quickly as will nitrogen. Let us prove it. I will plunge the end of the bent tube of our apparatus into the bowl of water and fill with hydrogen either the gage or a tall bottle with a wide neck, proceeding exactly as I did with oxygen."

Accordingly the gage was filled, after which Uncle Paul continued:

"Here is our gage, now, full of hydrogen. I lift it out of the water."

With these words he took the gage by its foot and

withdrew it from the bowl, holding it upside down as one would to empty it of a liquid. This procedure seemed to the boys to betray absent-mindedness on their uncle's part.

"If you hold it that way," they exclaimed, "the gas will all get out. The mouth is pointing down, and it is n't corked."

"No, my lads, the hydrogen will not get out. It is much lighter than air, and so tends to rise and not to fall. To keep it from escaping, we must block its way above, not below, and this I do by holding the gage upside down. There being no outlet upward, the gas is held captive. As to the open mouth below, we need not give it a thought; our hydrogen cannot go down and get out that way. I put a lighted candle into the gage and push it up almost to the inverted bottom. See what happens. The lowest layer of hydrogen, being next to the outside air, immediately catches fire with a slight explosion, and the flame gradually works its way up to the top of the column of gas. But as for the candle-flame, it went out at the very first, being smothered by the hydrogen as quickly and completely as it would have been by nitrogen."

This seemed very strange to the boys; they wondered how a gas that burns so well itself could put out a fire already burning. But the explanation soon given by their uncle was found to be simple enough.

"All burning," said he, "let us repeat and continue to repeat until the mind is quite familiar with this first principle—all burning, I say, is nothing but the chemical combination of some substance with oxygen, which is always present in the air. Where there is no oxygen, there nothing will burn. Well, then, the candle, on being thrust into the gage of hydrogen, went out because it did not find there the gas necessary for feeding its flame; it found no oxygen, and the other gas was unable to take its place, although very inflammable in itself. This gas, the hydrogen, took fire, but at first only in the bottom layer, because there and only there, next to the outlet, was there any air to feed the flame. Then this flame worked slowly upward from bottom to top, as the consumed hydrogen gave place to the air crowding in from below.

"Hydrogen is about fourteen times as light as air. This has been ascertained by means of chemists' scales, which are so delicately poised as to tip under the weight of a hair. Although an extremely light gas, hydrogen still weighs something, about one decigram to the liter. No other substance even among the most subtle, the gases, weighs so little. A liter of water weighs a kilogram, or ten thousand times as much as hydrogen. The heaviest of known substances is a metal called platinum,[1] which weighs twenty-seven times as much as water, and consequently twenty-seven thousand times as much as hydrogen. Between these two extremes range all the other substances known to us, some being heavy and others light according to their

[1] A slight correction is necessary here. Platinum is now known to weigh twenty-one and one half times as much as water, and two other metals, iridium and osmium, are a very little heavier. These three metals are much alike and occur together.—*Translator.*

position in this scale. Accepting these statements as verified, we will confine ourselves to showing by experiment that hydrogen is indeed much lighter than air.

"You have just seen how the gage must be held,—that is, with its mouth downward, to keep the hydrogen in. On account of its extreme lightness this gas escapes upward. Hence we may keep it confined by interposing some obstacle to this upward flight. Now let us prove that under the opposite conditions it will escape. We hold the gage upright, its mouth at the top, as if we had to do with nitrogen, oxygen, or atmospheric air, all three of about the same weight. Nothing standing in its way above, the hydrogen will quickly escape, you may be sure."

Refilled with hydrogen, the gage was set upright on the table, and they waited a few minutes. Nothing could be seen to go out or to come in. The sharpest eye could not have detected the departure of one gas and its immediate replacement by another.

"We have waited long enough," announced Uncle Paul. "There cannot be any hydrogen left there now. It is gone, and air has taken its place."

"How do you know?" asked Emile. "For my part, I can't see that anything has happened."

"Nor I, either; and if we had only our three pairs of eyes to decide the matter, the gage would keep its secret and never tell us what has taken place. But a lighted candle will tell us what our eyes cannot. If it keeps on burning in the gage, it will show that the latter contains air; if, on the contrary, it goes out after setting fire to the contents, it will mean that hydrogen is present."

A lighted candle was lowered into the gage and continued to burn there the same as before, proving that the hydrogen was gone and air, a heavier gas, had taken its place.

"If we lowered an open can of oil into a barrel of water," Uncle Paul went on, "what would happen? The water, being heavier than the oil, would force the latter out of the can and take its place, while the oil, being lighter than water, would rise and float on the surface. That is the way air and hydrogen act when the gage is set upright. But I have a still better experiment to show you in proof that hydrogen is lighter than air. With a few straws and a little soap-suds we can give a fine demonstration of the lightness of hydrogen. This is the way of it. You know better than I what will happen if we wet the end of a straw in soapy water and then gently blow through the straw. Emile used to play at that game not so very long ago, and he found it great fun."

"You mean blowing soap-bubbles?" Emile was quick to rejoin. "Oh, that 's fine, Uncle! On the end of the straw there comes a bubble, and it swells out bigger and bigger, to the size of an apple or an orange if you blow it the right way. And you can see all the colors of the rainbow on it,—blue and green and red, and so on,—more beautiful than the finest flowers in our garden. You don't dare move the straw for fear the magnificent bubble will burst. But before long it does burst, anyway, all of a sudden, and you don't know where it 's gone to. How sorry I 've felt, many a time, because my soap-bubbles would n't fly up into the air and soar about with all their splendid colors!"

"You won't have any reason to be sorry this time, my child," his uncle assured him, "for you are going to see your bubbles soar upward in fine style, and that without any coaxing."

"My beautiful soap-bubbles?"

"Your beautiful soap-bubbles."

"Then I shall like them better than ever."

"Show us first some of the kind you know so well how to make."

Emile took a straw, thrust one end into the soapy water that had been prepared, withdrew it, and blew gently into the dry end, making a series of bubbles, the largest of about the size of one's fist. All, as the filmy envelop became thinner with continued blowing, showed the brilliant hues of the rainbow; but they also, as soon as detached from the straw, all fell softly to the floor. Not one would float aloft in the air.

"And why should they rise?" asked Uncle Paul. "They are filled with air that is no different from the air all about them, and consequently receive no impulse from this gas either to rise or to fall. But their covering, which is made of a thin film of soapy water, is heavier than air, and so causes them to fall. Hence, if we wish our bubbles to rise, we must fill them with a gas lighter than air, a gas that will by its lightness not only make up for the weight of the envelop, but will also lift it up and bear it skyward. That gas is hydrogen."

"But how can I fill my bubbles with that?" asked Emile. "I can't blow hydrogen into them with my mouth."

"We will make the bottle do the blowing, and it will blow as well as any of us. First I take out the bent tube and put in a straight one, running side by side with that used for pouring in the acid and extending upward a little farther. But, as the tube is rather large, I contrive a smaller outlet at the top by inserting a straw wrapped at the lower end in a bit of wet paper. It is at the upper end of this straw, whence issues a jet of hydrogen—or a breath of hydrogen, if you prefer to call it so—that the soap-bubbles will form. I place the bottle upright so that the little balloons, borne upward by the lightness of the gas, will find nothing in their way. Now all we need do is to take a wisp of paper or something else and from time to time put a drop of the soap-suds on the end of the straw, whereupon we shall see bubbles form, filled with hydrogen."

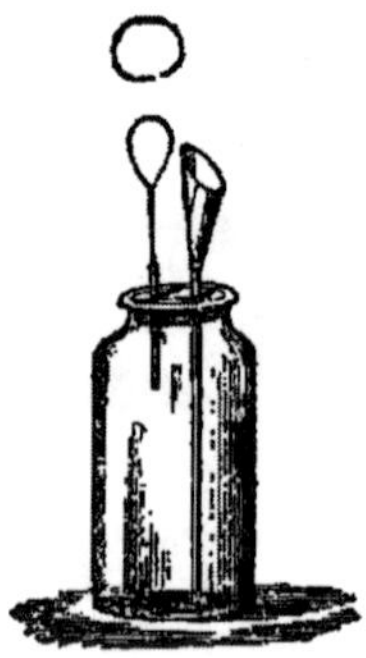

No sooner said than done. At the end of the straw, which was kept supplied with soapy water, there appeared a succession of transparent globes, sometimes larger, sometimes smaller, but always in an upright position on the tip of the straw and straining to get away from it. Many succeeded as soon as they were big enough, and then away they soared, rising rapidly and soon reaching the ceiling of the room, where they burst on touching it. Others burst before they could get clear of their moorings. Not for a good deal would the boys have missed seeing this ravishing spectacle. With wondering gaze they followed each balloon every instant, from start to finish. First they beheld it as a tiny bubble, then steadily swelling and all aglow with brilliant colors. It would sway a little on the end of the straw, then tear itself away, and off it would go in its flight to the ceiling. Oh, how gracefully it rose! But all too soon the ceiling was reached and the magnificent sphere shattered. Another soon followed it, however, and then another, and still another, as many as one chose. Jules was thoughtful, Emile jubilant.

"I am going to make this chemical diversion of yours still more entertaining," said Uncle Paul. "Tie a bit of candle to a long stick, light the candle, and then hold it up to a bubble while it is in the air."

Emile was not slow to carry out these instructions. Fastening a candle to a reed, he gave chase after one of the bubbles as it rose. *Flack!* went the little balloon, there was a sudden burst of flame in mid-air, and then nothing more; the whole thing had vanished. Emile gave a start; he had not expected so sudden a flash or one of so short duration.

"Does that startle you?" asked his uncle. "Did n't you know that hydrogen is exceedingly inflammable? Touch a lighted candle to a bubble filled with this gas, and there can't fail to be an instantaneous outburst of flame. That is the whole secret of these little aërial fireworks so astonishing to you."

"Yes, it 's simple enough, but I was n't expecting it."

"Now that you know what to expect, let us try it again."

The experiment was repeated several times, Emile allowing the bubbles to rise half-way to the ceiling and then touching them off with the candle. Not one of them, however quickly it rose, escaped the alert incendiary's pursuit. Thus it was shown in a highly diverting manner how readily hydrogen catches fire. Jules, who never asked idle questions, finally broke his silence,

"Our soap-bubbles," said he, "hit against the ceiling and that 's the end of them. Would they go up very high if they had plenty of room? Where would they go to?"

"In the open air and with nothing in the way they might rise to a great height if they did n't burst too soon; but the least agitation, the slightest breath of air, is sufficient to destroy them, so delicate and fragile a thing is a soap-bubble. Nevertheless if the atmosphere is very calm, the bubbles may last long enough to soar out of sight. We can try the thing out of doors this minute, for luckily the air is perfectly calm just now. Not a leaf is stirring on the tree in the garden."

The apparatus was carried outside and the bubble-blowing resumed. Many of the bubbles burst when no higher than the roof of the house, while others, though very few, rose out of sight. In a short time even Emile's sharp eyes could no longer distinguish them against the blue sky.

"Do they go very high?" asked Jules.

"I think not. A hundred meters, more or less, but at that height, being so small and so transparent, they become invisible. Their extremely thin and delicate covering, too, bursts before long. The one you are looking for now up there, hoping to catch another glimpse of it, is probably no longer in existence."

"But if the covering never broke, how high would the bubbles go?"

"On that point I can speak rather more definitely. Learned men who wish to explore the upper regions of the atmosphere and find out what is going on there, make enormous balloons of some strong fabric, and then varnish them outside and fill them with hydrogen, just as we are now filling our soap-bubbles. With these durable balloons they can go up to any height they please. The most daring have gone up ten thousand meters."

"Why not higher?" asked Jules. "I 'd have gone a good deal higher if I 'd been in their place. I should have wanted to see what there is at the very top of the blue sky. How beautiful it must be up there above the clouds!"

"In their place you would have done as they did, my dear boy, or probably not so much, for it takes almost superhuman courage to dare to visit those high regions. When you get to where there is not air enough, breathing becomes impossible and you have to come down in a hurry, or you are dead in a few minutes. That is why, up to the present time, the greatest height attained by man is about ten thousand meters."

"But the hydrogen balloon could go still higher if there was no danger for the balloonist?"

"Without a doubt, much higher."

"How high?"

"I can't tell you exactly, but let us say twice as high. All that I can be sure of is that balloon-ascent has some fixed limit, no matter how light and how skilfully constructed the balloon may be. The layer of atmosphere is thought to be only about fifteen leagues in thickness. Nothing that rises from the earth by reason of its superior lightness to air can pass that limit, since it no longer has air to buoy it up. At about fifteen leagues above the earth's surface, therefore, the ascent of any substance, even though it be the lightest gas, ceases."

"I 'd be satisfied with ten thousand meters, or even much less, if I only had a strong enough covering to my balloon not to break at a mere nothing, as our soap-bubbles do."

"You shall have your strongly covered balloon no later than to-morrow."

"And can I send it up as high as I want to?"

"Yes, as high as you want to."

At this prospect Emile clapped his hands with delight, while a gentle smile of satisfaction showed that Jules, too, was not indifferent to their uncle's promise. If he could not, himself, explore that beautiful blue void whose mystery so fascinated him, he could at least send a hydrogen balloon up there.

"One more question," said he, "before we stop blowing soap-bubbles. When they are filled with our breath they have all sorts of bright colors, and when they are filled with hydrogen they look just the same, so it is n't what 's in the bubbles that makes those beautiful colors?"

"No, my boy; those colors, which are the same as you see in the rainbow, do not come from air or hydrogen, nor do they come from the soapy water that makes the outside of the bubble. They are simply the play of light on the extremely thin covering. Whenever any transparent substance, whatever its nature, is in the form of an exceedingly thin film, the light striking on it causes this splendid coloring. Put a drop of oil, for instance, on still water, and the drop will spread out in the thinnest layer imaginable, whereupon the rich colors you speak of will appear. A soap-bubble or a thin layer of oil or a film of any transparent substance is called iridescent because it shows the colors of the rainbow, which the ancients called *iris.*"

## CHAPTER XIX

### A DROP OF WATER

"I PROMISED," Uncle Paul resumed, "to show you a balloon with so strong a covering that, when inflated with hydrogen, it could rise without danger of soon breaking. The time has come to keep my promise. Emile ought to have among his old playthings just what we need for our purpose. He doubtless remembers the pretty red balloons he used to buy at two sous apiece. To hold one of those balloons by its long string was an unfailing delight to him."

"Oh, yes, I remember them," Emile was quick to rejoin. "They were more fun than my Noah's ark with all the animals arranged in pairs, or my lead soldiers that I used to set up in companies on the table. But they are all crumpled up now in my old toy-box, and they 've been there a long time, ever since the day after I got them; for they would n't go up at all after the first day, but fell right down to the ground, though they were all right for a little while and went up finely."

"And have you never wondered why those little red balloons, so ready to go up the first day, were so very slow about it the next day?"

"Oh, yes, I 've often wondered, but I could n't tell why it was so."

"I will tell you the reason. Those balloons were filled with hydrogen, the same gas that we used yesterday for the soap-bubbles. Their covering is a fine membrane of rubber, very elastic, so that it stretches freely under the pressure of the hydrogen inside. Despite its extreme thinness such a covering is, so far as we can see, air-tight, being far superior in this respect to even the most closely woven fabric with its countless little open spaces between the meshes; and yet so subtle a gas is hydrogen that it manages to get through and make its escape. Then little by little the balloon collapses; or else, remaining full and round, it exchanges its hydrogen for air, the two gases passing through the rubber envelop in opposite directions. Consequently, whether by partial deflation or by exchange of hydrogen for air, the balloon loses much of its buoyancy before long, and after twenty-four hours cannot rise at all. To give it new life, we must again fill it with hydrogen."

"Oh, if I 'd only known that before, you 'd have had me teasing you to put new life into my balloons."

"A little chemistry has its advantages, has it not, my boy, were it only to put new life into toy balloons that are dead? If yours are still in good condition —I mean if there are no holes in the rubber— nothing is easier than to make them as buoyant as ever. Go and get them."

Emile ran out of the room and was back in a trice with two little red balloons, shapeless and shriveled, with not a breath of hydrogen in them;

nor had there been for a good while. But the child took such care of his things, was so particular about putting them away properly, that all his old toys were still in excellent condition. The rubber membranes that had once been lively little balloons, but were now all lax and lifeless, had not the slightest hole in them, as Uncle Paul soon learned by blowing into them.

"Our little balloons are all right," said he, "and now to work! I take a common bottle, the first to come to hand, holding about a liter, and into it I put some water and a good handful of zinc scraps. Then through a perforated cork stopper, fitting the neck closely, I pass a straight glass tube or, if that is lacking, a pipe-stem, or, better still, a goose-quill. The free end of this I next insert in the opening of the balloon and bind it with a string to make everything tight and prevent the escape of gas. Now I pour sulphuric acid into the bottle, and when the bubbling of the mixture is well under way, and the hydrogen is being released in ample quantity, I squeeze the balloon with my hand to drive out any air through the goose-quill, and still squeezing it I push the cork firmly into place. That done, I let go of the balloon and allow things to take their course. The crumpled membrane receives the gas, swells up gradually, and is stretched taut by the pressure of the hydrogen discharged from the bottle. Now you see it all spherical, and threatening to burst if I let much more gas go in. Finally, with a strong thread, I tie tightly the mouth of the balloon a little above the goose-quill to which it is attached. In that way the imprisoned hydrogen is retained, for a while at least. I then free the balloon from the goose-quill, so that the hydrogen forming in the bottle may escape and not accumulate to the point of blowing out the stopper with an awkward splash of the liquid itself."

"Now let 's see if it will go way up high," proposed Jules as the balloon swayed this way and that, straining to get free, but held back by his uncle's detaining hand.

"I don't want to lose it," said Emile. "Let 's tie a string to it. Here 's a long one that will do."

"Before talking about strings," was his uncle's advice, "let us consider the matter a little. How much gas does our little balloon hold? A liter at the most. Conseqently, the weight of the hydrogen that fills it is about a decigram. The same volume of air weighs fourteen times as much, making the difference between the two about thirteen decigrams. We will say the rubber covering weighs a gram, or ten decigrams. Three decigrams, then, are left as the measure of our little balloon's upward tug on its string. And with such feeble power at its disposal you would have our balloon drag the weight of a long string. Why not a rope or a ship's cable?"

"You are right. The balloon is n't strong enough to lift a long string up into the air. Then suppose we take a fine thread."

Attached to the end of a thread of this sort, the balloon rose, but, disappointing the children's expectations, it refused to go very high.

"It has stopped going up," said they. "Why?"

"Because the higher it rises the more thread it has to pull up after it, and the weight of this is added to that of the balloon itself. So the time soon comes when the weight of the rubber covering, the hydrogen, and the lifted thread all together equal the weight of the air displaced by the hydrogen; and from that moment further ascent becomes impossible. As Emile wishes to keep this balloon, we will inflate another and let it go without a thread to weigh it down. Free from this, it will go up as high as you wish."

And in fact, set free with no thread to shorten its flight by increased weight, the little balloon rises quickly and is very soon out of sight. How high will it go? Who knows? But whatever height it attains, sooner or later it will come down, because through the fine rubber covering there is taking place a constant exchange of gases, hydrogen going out and air coming in. Thus becoming always heavier, the balloon gradually falls back to the ground. But meanwhile who can say whither the winds may have wafted it?

"If Emile had n't kept the red balloons that we used to play with, could n't we have used a pig's bladder instead?" asked Jules. "That would give us a balloon all ready made, easy to get, and of a good size."

"I should have tried the pig's bladder only in case I had been unable to get anything better. It would have given us a large balloon, it is true, and made of a strong membrane; but it is loaded here and there with a layer of fat, which would be objectionable. You must not forget that the covering of our balloon should be as light as possible, so that the hydrogen may not lose its power of flight by having too great a load to carry. A liter of this gas cannot raise much more than a gram. Let us say the bladder will hold four liters, making about four grams that the contained hydrogen can raise. If, then, the bladder exceeds this weight, the balloon cannot rise. Consequently, if we propose to send up a balloon of this sort, the membrane must first be rid of its layer of fat, scraped down, its thickness reduced here and there, and all superfluous material removed, so as to lessen the weight as much as possible, while at the same time care must be taken not to make any holes in the membrane. With these precautions, which call for the utmost patience and skill, success is possible.

"When our village festival is held, one of its features is the ascent of a balloon, filled, not with hydrogen, but with hot air. It goes up amid cheers from the crowd and salvos from the village artillery, consisting of mortars loaded with charges of powder well rammed home. Why should not we, too, in our chemical pastime, have a discharge of artillery to accompany the ascent of our balloons? A bottle shall be the mortar, hydrogen the powder. Every time we set fire to hydrogen, even to a tiny bubble of it rising to the surface of the water, you heard a slight explosion. It is only necessary to find out what causes this explosion, and then to produce a louder one. It is presumable that air is necessary here, or more accurately the oxygen contained in air. Let us see, then, how hydrogen and air behave when mixed together.

"Into this narrow-necked bottle holding about a quarter of a liter, I pour enough water to fill it one third full, and then turn it upside down in the bowl so that it may receive a supply of hydrogen, using the bent tube once more and starting the release of the gas with sulphuric acid as before. Air fills two thirds of the bottle, water the other third. Hydrogen from our apparatus expels this water and takes its place. Thus the bottle contains a mixture of air and hydrogen, two parts of the former to one of the latter. I cork it tightly and wrap it up in several folds of a towel, leaving the neck free. Our gun is loaded. Now to fire it off."

With these words Uncle Paul grasped the towel-wrapped bottle by its middle with one hand, uncorked it, and held its mouth to the flame of a candle burning on the table. An explosion followed, and it was loud enough to make both the children start.

"Hurrah for the little hydrogen pistol!" cried Emile. "Famous gunpowder that makes. Once more, Uncle, please!"

The bottle-pistol was fired off again charged in the same manner as before,—that is to say, filled with the invisible powder of mixed hydrogen and air. Uncle Paul repeated the process as many times as was desired, and the explosions were more or less violent according to the proportions of hydrogen and air used. Some were full-toned and short, like the discharge of a firearm, others were only a noisy blowing, and others still, to Emile's unfailing amusement, sounded like the yelping of a little dog that has had its tail stepped on.

"My artillery," resumed Uncle Paul, "shows you that hydrogen and air form an explosive mixture that will take fire immediately at the touch of a flame and will explode violently. This mixture, invisible though it is, has no lack of power, and could blow its container to pieces if denied a sufficient outlet. That is why I wrap the bottle in a towel to arrest the fragments if the rupture should occur; and for a like reason it is well in these experiments to select a bottle of rather small size, a quarter of a liter at most. With a larger one there would be serious risk of injury: the gun might burst and wound the gunner.

"Air, as you know, is a mixture of an active gas, oxygen, and an inactive one, nitrogen. In exploding hydrogen it is clear that nitrogen plays no part; or, rather, by its inertia and considerable volume it impedes chemical action and muffles the detonation. The oxygen alone is active. Let us, then, do away with the nitrogen and use pure oxygen, and the detonation will be much louder. What we need for this purpose is all ready here, as I took care this morning to prepare a bottle of oxygen beforehand. It stands there in the corner, upside down, with its mouth in a glass of water. Before going farther I must tell you that to get the loudest explosion, hydrogen and oxygen should be mixed in the proportion of two parts of the former to one of the latter.

"I fill with water a wide-mouthed glass jar to

serve as receiver for holding our explosive mixture. Turning it upside down with its mouth in the water, I discharge into it one part of oxygen, using the first bottle at hand as a measure. Then I add two parts or bottlefuls of hydrogen. That done, our gunpowder is ready. Look into the jar. What do you see there? Nothing. Nevertheless it holds a dangerous explosive which it would be unwise to set fire to without proper precautions. The glass jar would fly into fragments that might seriously disfigure us. But there is nothing really to be feared as long as no match is applied. If you ever undertake experiments of this sort by yourselves, bear in mind that the mere fact of having water at hand does not insure you against the serious results of carelessness. This explosive does not in the least mind getting wet; it could stay next to water indefinitely without losing any of its terrible explosive power. Dry and wet are all the same to it.

"With the aid of a funnel, I fill with the gaseous mixture the bottle we have just been using, corking it and wrapping it carefully in a cloth to guard against a rupture, a thing more to be feared now than in our previous experiment. Now I have only to uncork my piece of artillery and hold its mouth near the candle-flame. Attention! Say 'Fire!'"

"Fire!" cried the boys in unison.

*Bang!* went the gun, and the room rang as if a rifle had been discharged. Emile gave a jump; indeed, he was almost frightened.

"To think that something you can't even see

should make such a noise!" he exclaimed. "I can hardly believe it. If I 'd known beforehand what was coming I should have stopped up my ears."

"Ah, indeed! That would look well, would n't it, Emile stopping up his ears so as not to hear the chemical pistol go off! No flinching, my boy, or I stop firing."

The candle having been relighted—for it was blown out with each burst of ignited gas from the bottle—Uncle Paul repeated the experiment. The explosion made the window-panes rattle; but this time, to make amends for his former show of fear, Emile stood as firm as a rock. He even forced himself to watch the proceedings with unwavering attention, and he saw a tongue of flame almost a meter long dart out with furious impetuosity from the mouth of the bottle. After a few more repetitions of this performance he became so hardened that he asked the great favor of being allowed to hold the gun in his own hand and fire it off as his uncle had done.

"I grant your request most willingly," was the reply. "There is nothing further to fear now. The bottle has been well tested, it has withstood the strain put upon it, and it will continue to do so. However, as an extra precaution keep it securely wrapped in the cloth, so that if by any chance it should burst the pieces would still stay in the wrapping. Grasp it boldly in your hand; there is nothing to be afraid of."

Without flinching, Emile managed the bottle when it had been charged by his uncle. With sober face and standing rigidly erect, like an artilleryman discharging his cannon, he fired off the glass gun. Jules then took his turn, his delicate hand trembling a little with excitement. And so they alternated until the explosive gas was all used up. These repeated discharges made the rest of the family wonder what in the world Uncle Paul and his nephews could be up to with their chemicals and their bottles and their other pieces of apparatus.

"Now that our gun is silenced for want of ammunition," said Uncle Paul, "let us see what we have left from all this burning of hydrogen in oxygen. The two gases combine when the explosion takes place, and this combination is attended by a streak of flame, not very bright, which you saw shoot from the bottle. A new substance, a compound of hydrogen and oxygen, is formed as the explosion takes place, but it cannot be seen, as it is an invisible vapor. This must be condensed if we wish to examine it. But instead of igniting a great volume of gas all at once, which would be somewhat dangerous and, besides, would not serve the purpose of careful study, we will arrange matters so that our hydrogen and oxygen shall combine a little at a time; that is, we will light a jet of hydrogen and let it burn in the air so that we can watch it closely.

"Let us begin with the preparation of our apparatus. It will be the same as we used for blowing our soap-bubbles, except that the straight tube having a straw in its upper end will be replaced by another with a tapering tip. The channel as it is now would not do, being too large; it must be made smaller until the outlet does not much exceed the size of a pin. This is how it is done. A piece of glass tubing, as easy to melt as possible, is held in the flame of an alcohol lamp and twirled between the the fingers so as to be heated equally all around. When it seems soft enough it is pulled so that the dough-like heated part stretches into a thread and two sections are obtained with a mere string of glass between them. With one sharp tap in the middle of this connection we get two tapering tubes, either one of which will serve our purpose. Accordingly, we select one and thrust its blunt end into one of the two holes in the jar's stopper. The second hole receives, as before, the straight tube for the introduction of the acid. I will add that a clay pipe-stem could be used if necessary instead of the tapering glass tube. Of course, if you have a two-necked jar, one neck would receive the tapering tube and the other the tube for the acid.

"Water, zinc, and sulphuric acid having been put into the jar, hydrogen is released and flows in a jet through the tiny outlet of the tapering tube. I propose to light this jet, but here we must exercise a little prudence. We have just seen that hydrogen and air make an explosive mixture. Now, when the release of hydrogen began, the jar contained air, and hence an explosive mixture is inevitable at the start. If we were so careless as to hold a lighted match to the gas jet at this stage of operations, the dangerous mixture would explode in the jar, which might not withstand the shock; it might be blown to bits, or at the very least the cork would pop out and with it would come splashes of the acid, staining our clothes with red spots and, much worse, perhaps getting into our eyes. I warn you to beware of this explosive mixture and to be constantly on your guard if you are preparing hydrogen by yourselves. You must watch every moment to see that no air mixes with the gas you are intending to light.

"In the experiment before us the presence of air is inevitable at the start. What, then, is to be done? The release of gas must be allowed to go on for a while. As it comes out of the jar, the hydrogen brings air along with it; and, as this is not renewed, a time comes when there is none or next to none left. But as there is nothing to indicate when this moment has arrived, we must simply wait a while, and our waiting should be rather too long than too short, so as to assure us of safety. Here patience is prudence."

They waited while the release of hydrogen went on and the liberated gas came whistling through the tiny outlet of the tube. After a few minutes it was thought safe to proceed.

"That ought to do," declared Uncle Paul. "There can hardly be any more air left in the jar by this time. But we may be mistaken, and so, to prevent accident, I wrap the jar in a towel to arrest, if necessary, any flying fragments or splashes. With this final precaution, I hold a lighted paper to the hydrogen jet. Instantly the gas takes fire and burns quietly with a very pale yellow flame. All danger is past. As there was no explosion when I lighted the gas, there will be none. All the air is driven out of the jar, only hydrogen issues from the tapering tube. Our towel was not needed, but it should nevertheless be used whenever we suspect there may be an explosive mixture in the jar. Not to hide from view what is going on there, I now remove the protecting cloth, which is of no further use. Instead of worrying about an explosion that cannot occur, let us examine our hydrogen flame closely.

"At the tip of the tapering tube we see a steadily burning flame, fed by the gas constantly coming out; and this flame is of a very pale yellow and gives scarcely any light. Such is the appearance of burning hydrogen, of which until now you have caught only fleeting glimpses. It is not a bright flame, but an exceedingly hot one. Try it."

The boys held each a finger-tip near the paltry little flame, and each drew it back quickly. The heat was unbearable.

"Oh, oh!" Emile cried out with pain. "That flame does n't look like much, but it 's a good heater, all the same."

"It ought to be, coming as it does from burning hydrogen, the best of fuels. Do you remember what the blacksmith showed us?"

"Do you mean his wetting the coal in his forge so as to heat his iron white-hot?"

"Yes. Water, decomposed by burning coal, gives hydrogen; and this hydrogen burns and thus adds considerably to the heat."

"Then we could make iron red-hot in this flame, though it is so small and pale?"

"You could make it not only red-hot, but white-hot. Look here. I hold the end of an iron wire in the flame, and immediately it becomes dazzling to the eye. It is thus that the blacksmith's iron bar is heated in the forge when he wets his coal.

"Another peculiarity, less important but more curious, is that the hydrogen flame sings. Yes, my little friends, it sings, and you shall hear its song as soon as I have provided it with a suitable musical instrument. This instrument is a glass tube about as long and as slender as a walking-stick. But the tube may be shorter and wider, with corresponding changes in the tone, deeper if the tube is wide, on a higher note if it is narrow. For want of such an instrument a lamp chimney will serve, or a pasteboard or even paper cylinder. They are made in several lengths and calibers. I have prepared a number of these tubes, one of which is of glass. I begin with that."

Uncle Paul, holding the glass tube upright, lowered it so as to enclose the flame, whereupon there was heard a musical sound, continuous and full, not unlike that of an organ-pipe. According as the tube was lowered or raised, causing a greater or less extent of flame to be within it, the note was on one pitch or another, changing abruptly from octave to octave in a manner to jar upon the least sensitive ear. Sometimes the note was tremulous, then uniform, and then tremulous again. Occasionally the effect was that of a solemn prayer muttered in the hushed seclusion of some chapel, after which would come a response in high falsetto. In short, whatever the note, it always had a somewhat strident tone, affecting the ear as a disagreeable buzzing. By repeatedly changing the tube, selecting now a long one and now a short, now one of greater caliber and now one of less, sometimes one of paper, then of glass or metal or pasteboard, Uncle Paul ran through the entire gamut in a riot of ear-piercing notes.

"Oh, what a crazy medley!" exclaimed the young hearers, overcome with laughter and unable to stand the discord any longer. They sought relief by stopping up their ears. "Oh, if Bull were only here! He barks when any one plays the flute or violin, and what fun it would be to hear him join in with this hydrogen orchestra! Let 's go and find him."

The dog was soon found, and he came readily enough, perhaps thinking he was going to get a bone. At the first note of the mad music the animal became excited and began to howl and moan, voicing his bewildered surprise in the strangest kind of sounds, all to the fit accompaniment of the hydrogen refrain. Emile and Jules burst out laughing at this instrumental and vocal concert, and even their sedate uncle failed to preserve his usual gravity.

"Put that disorderly pupil out of the class room immediately," he commanded, "or I shall be joining in your foolish merriment and the lesson on hydrogen will go unfinished."

When the dog had been banished and the hilarity calmed, Uncle Paul continued:

"You will of course understand that it was not

for the sole purpose of treating you to such a charivari that I took it into my head to enclose the hydrogen flame in a tube; there was back of this mad concert a serious motive, which I will explain after answering a question that both of you must surely have on your tongue's end. What makes the hydrogen flame sing? Enclosed in the tube, the gas jet meets with air, and so there is continually being formed an explosive mixture that causes a succession of tiny detonations, one immediately after another, making the column of air in the tube vibrate. The sound we hear is due to this vibration.

"But now let us drop this subject and see what is produced when hydrogen is burned. What becomes of the burned hydrogen? I take the glass tube again, and wipe it thoroughly inside with a wad of blotting-paper on a stick. So now we have it in a perfectly clean condition, without a trace of moisture on the inside. Again I pass it over the hydrogen flame. But don't listen now to the singing of the gas; watch what is going on in the tube. A fine dew soon forms on the inner surface of the glass, increasing gradually until it trickles down in colorless drops. This liquid is the burned hydrogen, the compound resulting from the combining of hydrogen with the oxygen of the atmosphere. From its appearance one would call it water, but before we can be sure we must taste it.

"With the tube I am using it is hardly possible to wet the finger-tip in the drops running down the inside; so let us make a slight change, substituting for the tube a large jar with a wide mouth. I wipe the inside carefully and then introduce the hydrogen flame. Dew again appears, and drops collect and run down. If we wait long enough, some of them will reach the mouth and you can wet the tip of a finger in them."

After the flame had burned some time in the jar, which was held upside down and, to enable the condensed liquid to gather in one place, a little slanting, some drops did in fact collect at one point on the mouth of the jar. At their uncle's bidding the boys hastened to wet a finger-tip in the liquid and taste it.

"It hasn't any taste," declared Jules, "or any smell or color. I should almost say it was water."

"You may say so without the 'almost,' for that is just what it is. This is the wonderful thing I wished to teach you when I made the flame sing. Water is burnt hydrogen; it is composed of hydrogen and oxygen. Commonly looked upon as the direct opposite of fire, water in reality combines the elements essential to the hottest kind of a fire, namely, hydrogen, the best of fuels, and oxygen, in which metals, even including iron, will burn. The two gases are present in unequal parts, one of oxygen to two of hydrogen. That is why, when I prepared the mixture that exploded so loudly, I put into the bottle two measures of hydrogen and one of oxygen. The explosion of this mixture produced a little water, which, vaporized by the high temperature, rushed with much violence and noise out of the bottle. From the loudness of the explosion, causing the window-panes to rattle at each occurrence, you might think a good deal of water had been produced. Undeceive yourselves; great as was the noise, the quantity of water produced was small, very small, perhaps a single drop. You shall judge for yourselves from the figures given by chemistry. It tells us that to make one liter of water we must have one thousand, eight hundred, and sixty liters of our explosive mixture, of which six hundred and twenty are oxygen and one thousand, two hundred, and forty, or twice that amount, hydrogen. How much water, then, would come from our little bottle holding a quarter of a liter? Hardly any at all. What a noisy celebration over a chemical marriage from which is to be born a single drop of water!

"Now we can see the reason for using sulphuric acid with our metal, be it zinc or iron, in decomposing water. We know that an acid unites with metal if the latter is first converted into an oxid. A salt is formed by this union of acid and oxid. Zinc, sulphuric acid, and water are put into a jar together. The acid has a natural liking for the metal if the metal first becomes an oxid; and the metal, on its part, has a liking for the oxygen. The combined effect of these two likings is the decomposition of the water, which lets its hydrogen escape and gives its oxygen to the zinc. Thus the metal burns at the expense of the water, which furnishes the element without which nothing can burn. It burns, I say, or in other words it becomes an oxid, which immediately combines with the sulphuric acid to make a salt called sulphate of zinc. In short, what zinc undergoes with the help of water is exactly what it undergoes in a brazier when it burns with the help of air. The metal burns there, or is turned into an oxid. It is this burning of the metal that produces the heat generated by our hydrogen apparatus when it is in operation. You were surprised that a liquid should get hot without fire. Now the mystery is explained.

"I said the zinc was converted into a salt. This salt, this sulphate of zinc, readily dissolves in the water used in the experiment and only slightly diminished in volume by the decomposition of a very small part of it. The metal, then, disappears from sight just as sugar in melting disappears. Let us now take a look at our hydrogen apparatus. For some time it has been producing no gas. The zinc is all gone except a few blackish flakes resulting from impurities in the metal. It is dissolved in the liquid, having first been turned into a salt, and the liquid itself remains as colorless as at first. We will let the jar stand in a corner, and little by little the dissolved substance will crystallize out, and we shall have a white deposit of an unbearably sharp taste. This white substance will be sulphate of zinc."

## CHAPTER XX

### A PIECE OF CHALK

"TO-DAY, my young friends, we shall have no thundering artillery, no furious outbursts of flame, no ear-splitting concert on pipes of glass and pasteboard, no boisterous festival to celebrate the birth of a drop of water. But the lesson will be none the less important because it is quiet. We are going to inquire what becomes of coal, or carbon, when it burns. We have seen how brilliantly it burns in oxygen, and we shall not soon forget that magnificent display. In the process of burning, an invisible gas is produced, a gas that we called carbonic acid, knowing it to be an acid because, like other acids, it reddens, even though but slightly, the blue of litmus. This gas, known to us so far only by name, merits closer study. First of all I will show you how to recognize it, whatever may be its origin.

"Here is a piece of lime. I sprinkle water on it to slake it, as the masons say, and it turns hot and steaming and crumbles to powder. I add more water, much more, so as to make a thin paste. You have not forgotten that lime is slightly soluble in water, and it is a solution of that sort I desire, one that is perfectly clear and without a trace of undissolved lime. This I obtain by straining the paste through a paper filter placed in the mouth of a funnel. You know that when we wish to separate fine grains from coarse in a mixture of the two, we use a sieve that allows the former to go through while it retains the latter. Well, the filter-paper is also a kind of sieve, all perforated with invisible pores that give egress to matter reduced to exceedingly fine particles by the act of dissolving, and bar the way to what is not dissolved. Druggists and grocers sell this filter-paper, which is much used in straining turbid wine and vinegar, or any other liquid that is to be freed of its dregs. It comes in rounds, larger or smaller as may be desired. By folding it over once we get a semicircle, by folding it a second time we get a quarter circle, another fold gives an eighth of a circle, and so on until it can no longer be easily folded. Then, by partly spreading out our folded disk, we obtain a sort of crinkled funnel, which we place in a chemist's glass funnel as a support, if such a funnel is at hand; otherwise a common tin funnel will do. A bottle for holding the filtered liquid receives the tube of the funnel.

"My filter is ready, and through it I strain the lime paste. Notice how thick and muddy the liquid is above the filter, and how clear and limpid below. What has gone through into the bottle looks like the very purest of water. Is not this round of paper a wonderful sieve in its perfect separation of the dissolved from the undissolved lime? The liquid that passes through is something more than water, clear though it is to the eye; it contains a little dissolved lime, as we find by tasting it. It is called lime-water, and we shall use it as a test for carbonic acid.

"We must now obtain some of this carbonic-acid gas by burning charcoal either in pure oxygen or in the mixture of oxygen and nitrogen that we call atmospheric air. We will adopt this latter method as being both simpler and quicker. Here are two bottles of the same size and both full of air. Leaving one for the present, I introduce into the other a piece of charcoal well on fire, and keep it there until it goes out, which it soon does. A little carbonic acid has formed. It is there in the bottle, invisible and mixed with what is left of the air originally in the bottle. The lime-water will betray its presence. I pour a spoonful or two into the bottle and shake it up, whereupon it loses its clearness and turns clouded and white. Is it really the carbonic acid that has caused this sudden change? In the bottle there is still some air, containing nitrogen and the little oxygen not used up by the burning of the charcoal. Could not one or the other of these gases have caused the milky appearance of the lime-water? We must make sure before asserting anything. So I pour lime-water into the bottle that has had no charcoal burned in it, and I shake it up, but with no result: it remains perfectly clear. Hence it must be the carbonic acid alone that turns the lime-water white, and neither oxygen nor nitrogen has anything to do with it. I will add that no gas except carbonic acid has this power of giving a whitish appearance to lime-water.

"So, then, we have in this liquid a useful means for distinguishing carbonic acid from all other gases. Suppose, for example, a certain bottle contains a gas that we know nothing about, and we wonder whether it may possibly be carbonic acid. Lime-water settles the question at once by either turning white when shaken up in the bottle or by remaining unaffected. Often when there has been some combustion of carbon without our knowledge, lime-water will decide the matter, its testimony being unimpeachable even where the presence of carbonic acid is held to be impossible. We shall have occasion to invoke this testimony in cases of the utmost importance. Let us therefore bear this truth well in mind: carbonic acid turns lime-water white and, conversely, whenever lime-water is turned white by a gas, this gas is carbonic acid.

"I pour into a glass the liquid whitened by the carbonic acid in the bottle. Hold it up to the light and look through it. What do you see? Whirling white flakes, fine particles like those in curdled milk. We let the liquid stand undisturbed a while, and the white particles slowly settle at the bottom, leaving the liquid clear once more. I throw away the liquid and keep the sediment. There is not much of it, hardly a pinch. What can it be? From its appearance one might say flour, starch, or powdered chalk. And chalk it is, surely enough, chalk like that used for writing on a blackboard.

"But don't for a moment imagine that the sticks of chalk used for writing are made in the same way as this pinch of white powder. They would cost an enormous sum if, in order to make them, charcoal had to be burned and then lime-water shaken up with the carbonic acid thus obtained. Common chalk occurs ready-made in nature; it only has to be separated from its impurities, pressed into a solid mass, and cut into sticks. Nevertheless we have just obtained real chalk by artificial means. And how has this come about? It is this way: the water contained lime in solution; carbonic acid came and took possession of this lime, combining with it and forming a salt known as carbonate of lime; and thus the white powder you see there is a salt, a compound of lime and carbonic acid,—in short, carbonate of lime.

"In its natural state this substance, though composed always of carbonic acid and lime, may be of very different degrees of fineness, consistency, and hardness. If it is rather soft and tends to crumble, it is chalk; if hard and coarse, limestone, building-stone, ashlar; if still harder and of fine, compact structure, it is marble. These various kinds of stony substance, so different in name, appearance, and use, are at bottom the same thing,—a union of lime and burnt carbon. Chemistry, which pays no heed to outer appearance, but considers only inner structure, gives them all alike the name of carbonate of lime. Hence we could if necessary get from chalk or marble or any piece of limestone a supply of carbonic acid that would not differ in the least from that obtained by burning charcoal.

"Now I beg you to note the curious lesson taught us by this lime-water. Suppose we wish to study the gas produced by burning charcoal. Do we heap up fuel in the fireplace and make a roaring fire? By no means. We need no mass of glowing charcoal, no fireplace, no furnace. A few bits of stone will furnish us the same gas, absolutely the same. Chemistry abounds in similar examples that upset our little stock of conventional notions. To the ignorant all this seems like magic. Are you seeking the best possible fuel? Chemistry tells you to look for it in water. Do you wish for some gas such as rises from burning coals? Chemistry informs you it is to be found in stone.

"There is carbon in chalk. The blackest of black substances enters into the composition of the very whitest. To doubt this would be impossible even for Emile, who is always demanding proof after proof, whenever any statement surprises him by its novelty. It was really carbon, in the form of charcoal, that I burned in the bottle; a gas called carbonic acid was thus produced; and then, but not before, the lime-water brought into contact with this carbonic acid showed tiny chalk-flakes whirling about in it. There is carbon, I say, in chalk; but it is in the burnt state, and to burn it over again is impossible unless its present partnership with oxygen is first dissolved. Consequently, chalk is incombustible. But carbon abounds in numerous other substances, and in an unburnt condition; therefore it is combustible. It cannot be seen with the eye in these substances, nor is there anything to make one suspect its presence. Take, for example, the material of which so-called wax candles are made: this is of a beautiful white appearance, and yet it contains carbon in abundance, as is proved by the black smoke rising from the burning wick. But, disregarding this smoke, let us examine the substance itself and find out whether it does really contain carbon. Our course is marked out for us in advance: we have only to light the candle and see whether there is any carbonic acid formed. If so, it is an incontestable proof that there is carbon in the white substance of the candle. Let us try it.

"I renew the air in the bottle by filling it with pure water and then emptying it, as I have already shown you. Then I lower into the bottle a lighted candle fastened to a wire, and I keep it there until it goes out. Has any carbonic acid formed? Our lime-water will tell us. I pour a little into the bottle and shake it up. Ha! Look there: the lime-water has taken on a cloudy whiteness, and hence we know the burning of the candle has produced carbonic acid, proving that the white substance of which the candle is made contains carbon. Nothing difficult about that.

"Let us take another example. Paper contains carbon, and we have but to burn a piece of it and examine the charred fragments, to see, from the black color, that it is there. But we will take no account of this indication, which, after all, might be deceptive, as all is not carbon that is black. Refilling our bottle with pure air, I twist up a good-sized piece of paper and burn it inside the bottle, taking care not to let the ashes fall, as they would interfere with the rest of the operation. That done, I call on the lime-water to give its testimony, and again the lime-water turns white. So there must be carbonic acid there, and hence carbon. It speaks for itself, you see.

"Once more: The black smoke of the burning paper and the black color of its charred fragments make us pretty sure, without the help of lime-water, that paper contains carbon, just as the black smoke rising from a smoldering candle-wick makes us think the candle has carbon in it; and this we infer despite the whiteness of both paper and candle. But now we come to a third substance that gives no similar indication of the presence of carbon. It is alcohol, or spirit of wine. Its strong winy odor proves at once that it is not water, though it is quite as clear. It takes fire readily and burns with a smokeless flame. Is there carbon in this inflammable and perfectly colorless liquid? Not a sign of anything black do we detect either in the flame or anywhere near it. There is no foul smoke, no dark-colored residue. Only lime-water can settle the question here. Into a small cup supported by a wire bent into a ring at one end I pour a little alcohol. I light it and lower it into the bottle, the air in which I have first renewed. As soon as the alcohol stops burning I apply the lime-water test. The water is whitened. That settles it. I can now affirm with absolute certainty that alcohol, a liquid as colorless and transparent as water, contains that compact black, opaque substance called carbon.

"In this way we might test a great variety of substances, and all in which the gaseous products of combustion produced a clouding of lime-water would be shown to have carbon in their composition. If I have laid some stress on this point, it is to show you by experiment the deceitfulness of appearances in determining the real nature of a compound substance. I have just proved to you, by facts that speak for themselves, the presence of carbon in substances that, from appearances only, would not be suspected of containing it; and thus you are prepared for the even more surprising demonstration that a piece of stone can be made to yield the gas we call carbonic acid.

"Chalk, marble, and all limestones contain carbonic acid, an acid of little strength and always ready to give up its place to any other acid that is more powerful. In chemistry the harsh law of the strongest prevails: get out of that and make room for me. If, then, we pour some strong acid on carbonate of lime, carbonic acid is set free, being ousted by the new-comer, which takes its place and forms with the lime a new salt. Sulphuric acid, for example, turns a carbonate into a sulphate, and phosphoric acid turns it into a phosphate. In both cases the carbonic acid is set free, and its release is accompanied by a foaming on the surface of the stone.

"The thing is worth seeing. Let us experiment with the pinch of white powder obtained when we shook a little lime-water in the bottle in which we had burned a piece of charcoal. Here is the powder in the bottom of this glass and not yet dry, but that will not interfere in the least with the success of our experiment. I let a drop of sulphuric acid fall on to the white paste. Immediately the mixture has the appearance of boiling and is covered with froth; it foams, as we say. This foam is made of the bubbles of carbonic acid driven out by the other acid. Now let us try some real chalk, such as is used for writing on blackboards. I take a stick of it and with the help of a slender glass rod I place a drop of sulphuric acid on the chalk. Where the acid touches the chalk there is a foaming appearance, a sure sign that carbonic acid is being driven out.

"You were already inclined to consider this white powder of the same nature as chalk, taking my word for it and also judging from the likeness of appearance between the two; and now a very decisive test proves it beyond a doubt. Both substances foam when touched with an acid, and both yield the same gas, as could easily be ascertained by operating on a larger scale and by collecting the bubbles of gas which cause the foam. The resemblance is not confined to mere appearances, but goes deeper and is fundamental; that is to say, the two substances tested are the same thing.

"Limestone, again, is the same thing. But how are we to know limestone when we see it, how distinguish it from other kinds of stone? These are questions to be answered without delay, as we are about to use this stone in obtaining a sufficient supply of carbonic acid for further experiments. We must not proceed in any haphazard fashion, trying the first stone we come to, and then another if that does not serve our purpose. We must be able at the very start to give our apparatus the right material for producing carbonic acid. Where is this material, this storehouse of gas, to be found, and how can we pick out our particular kind of stone from among so many different kinds? A drop of strong acid will give the answer. Here is a hard stone picked up in the bed of a brook. I touch it with a drop of sulphuric acid: no result, no foam; hence the stone contains no carbonic acid, is not a carbonate, and will be of no use to us in producing the gas we desire. We throw it away. Here is a second stone, as hard as the first. I test it in the same manner, and the acid makes it foam at the merest touch. So there is carbonic acid here. The stone is carbonate of lime, limestone. For one who is not well acquainted with the rocks of his region and cannot distinguish one from another by outward appearances, there is—as you now see, I hope—a very convenient way of settling all doubts as to what is limestone and what is not.''

''That is plain enough,'' Emile agreed; ''if it foams with a strong acid, it is limestone; if not, it is n't limestone. The one that foams will give carbonic acid, and the other won't, for the very good reason that it has none.''

''In our definition,'' Uncle Paul continued, ''I will change only one word, and I do this in order that our terms may be those used in chemistry. When I spoke of 'foaming' to indicate the bubbling or boiling produced when a gas is set free, I should more properly have said 'effervescing.' In chemical language, then, we say that limestone effervesces under the action of a strong acid, which is really the same as saying that it bubbles, foams, or froths; and a stone that does not effervesce is not limestone.

"The word 'limestone' means carbonate of lime; but there are many other carbonates, each metal giving one of its own, or sometimes more than one, as occasionally there are several carbonates for the same metal. Iron, copper, lead, zinc, to name no others, have each its own carbonate, just as calcium has, this last being, of course, carbonate of lime, or limestone. This carbonate is much more plentiful than any of the others, and plays a more important part in this world of ours; therefore I particularly call your attention to it. A good half of the soil is made of it. Great mountain-chains are blocks of this salt. Whether rare or abundant, all carbonates without exception have the peculiarity of effervescing when touched with an acid. Since they all contain carbonic acid, as otherwise they would not be carbonates, they all release this acid when a stronger acid comes to take its place. Effervescence or foaming is the invariable accompaniment of this release of gas. You will see before long what lessons are to be learned from this peculiarity.

"I put into this glass a pinch or two of ashes from the fireplace. If I were to ask you what these ashes contain, what would your answer be? You could not tell me, for neither sight, taste, nor smell gives us any information. But by a skilful roundabout method we can arrive at the answer, with evidence to back it up. I pour a little sulphuric acid on the ashes, and very decided effervescence follows, the mixture of acid and ashes foaming rather violently.

Consequently we infer— But who will tell me?"

"I know," Emile hastened to reply; "there is carbonate of lime in the ashes."

"I think," said Jules, "that Emile has jumped to a rather hasty conclusion. All carbonates effervesce with acid, and so the foaming shows only that there is a carbonate in the ashes, but does n't tell us which carbonate."

"You are right, my boy. The ashes contain a carbonate, but not that of lime; it is of another metal, potassium, a name you have already heard. If what I have just done to the ashes does not tell us the nature of the metal they contain, it at least tells us that in these ashes from our fireplace there is carbonic acid. In this connection it is well for you to learn that chemists determine the nature of a substance by just such tests as this. You hand to the chemist a rock, a mineral, an earthy substance, or anything else you deem worthy of serious examination. He tests the thing with one chemical and informs you that it contains iron; he tests it with another and says it contains copper; a third test proves the presence of sulphur; and so on. Yet neither iron, copper, nor sulphur was visible to the eye, nor did they become so in the course of the various tests. Nevertheless, that they are there is proved conclusively by the action of the various chemicals. When a piece of white marble effervesces on being touched with sulphuric acid, I conclude that the marble contains carbonic acid and, consequently, carbon. It is in similar manner that the chemist reasons from the tests he applies, ascertaining by these tests that one substance contains this and another that, without needing to see with his eyes either this or that.

"Now, then, let us get to work with our preparation of carbonic acid. For this purpose we have here a quantity of broken limestone. I put a good handful into a glass, and add water to moderate the action of the acid so as to give us a steady and not too rapid release of the gas instead of a violent effervescence, which would be difficult to manage. The acid I am going to use will not be sulphuric acid, which I used in the experiments you have just witnessed; and the reason for not using it is this: with sulphuric acid, carbonate of lime would change to sulphate of lime, or plaster of Paris, an insoluble substance that would encrust the fragments of stone and thus obstruct further action on the part of the acid, so that the release of gas would come to a standstill. The operation, though successful enough at first, would come to an untimely end. For an uninterrupted release of gas, the pieces of stone must keep their surfaces clean, not coated over with a protecting layer. In other words, the new compound formed with the acid must get out of the way as fast as it forms. This conditon is fulfilled if the new compound dissolves in the surrounding water, a result obtained by the use of hydrochloric acid."

"What acid did you say?" asked Emile.

"I said hydrochloric acid."

"What a funny name! I don't believe I can remember it."

"Don't blame me; I did n't invent it. We will call it, if you like, spirit of salt, the name it bears in workshops and factories where it is used. My friend the blacksmith, who is also a locksmith and a coppersmith, uses it to clean his old copper utensils, and he gave me this bottleful. Talk to him about spirit of salt, and he will know what you mean; call it hydrochloric acid, and he will give you a blank look."

"But why is it called spirit of salt? That 's a rather queer name too."

"It is so called because it is made from salt, common kitchen salt. As for the word 'spirit,' which seems to puzzle you in this connection, it is a remnant of the old-time language of early chemistry, which gave this name to all invisible substances, substances that we now call gases, in fact. To those chemists of long ago the invisible and inflammable vapor that rises from hot wine was the spirit of wine, and the pungent acid vapor derived from common salt treated in a certain way was spirit of salt. 'Spirit of wine' is a term still in common use, whereas 'spirit of salt' is heard only among artisans having occasion to use it.

"This acid has another noteworthy peculiarity: it contains no oxygen, is not a burnt metalloid as are sulphuric acid, carbonic acid, phosphoric acid, and the other acids we have had more or less occasion to speak of. It is made of chlorin and hydrogen, and this composition gives it the name you find so hard to remember. Chlorin, as I hope you have not forgotten, is the metalloid found in salt and in chloric acid. As to hydrogen, it is not necessary to remind you what that is.

"Briefly, hydrochloric acid or spirit of salt, whichever you please, is a yellowish liquid of a very sour taste, evaporating in the air in white fumes of an exceedingly pungent odor. I pour some into the glass containing water and pieces of limestone, whereupon there follows a lively boiling and bubbling caused by the release of carbonic acid from the stone and its replacement by the acid of salt; and this chemical action we shall turn to our account in the next lesson."

## CHAPTER XXI

### CARBONIC-ACID GAS

"YESTERDAY we found limestone to be rich in carbonic acid, and we also found that, to release this gas from the stone and obtain it by itself, all we need to do is to apply to the stone another and a stronger acid, preferably hydrochloric acid, a very cheap liquid and one that has the additional advantage of always keeping the surface of the stone clean. Our plan for to-day is to extract from limestone this gas of burnt carbon. The apparatus required will be the same as for the preparing of hydrogen,—that is to say, a two-necked bottle or jar, if our laboratory can furnish this handy utensil; or else, if our resources are more modest, a bottle with a wide neck to hold a large stopper having two holes running through it from top to bottom. Into one of these holes will be fitted a straight glass tube reaching to the lower part of the bottle, and through it, by means of a small glass funnel or, lacking that, a paper cone, the hydrochloric acid will be poured little by little so that the effervescence may not be rapid enough to make the foam rise too high and run over. Through the second opening will pass a bent tube to serve as conductor of the released gas.

"Here is what we are after,—a plain bottle with a large cork stopper having two holes through it. Into this bottle I put a handful of broken limestone of the hardest sort. If I had a piece of marble,—a fragment from some old bureau-top, for example,—things would go better; but, not having any, I do the best I can with common limestone, its only defect being a tendency to soil the liquid somewhat with its impurities. I add water and adjust the stopper with its two tubes. Of course the straight tube, not the bent one, is pushed down into the liquid. Now I pour in a little hydrochloric acid, and immediately we see a commotion, a kind of boiling, caused by the setting free of gas from the stone. From this point things will go of themselves, and we need only let the apparatus alone, confining our activities to the occasional addition of a little acid to keep the process from halting."

"Quick, quick, the bowl of water for collecting the gas!" cried Emile, seeing his uncle abandon the apparatus to itself in an apparently very careless fashion.

"The bowl of water is unnecessary here," his uncle assured him; "we shall get our carbonic-acid gas quite well without having to use that cumbrous thing."

"But the gas is just going to waste."

"We can afford to waste a little, so easy is it to make more and so inexpensive the process. What does it take to produce all we wish of this gas? Only a pennyworth of acid and a stone picked up by the roadside. Besides, there is a reason for my inattention to waste: there is air in the bottle, and I am allowing the carbonic acid to drive it out.

"Now it is done, or very nearly, and there is no more air left in the apparatus—except, perhaps, a mere trace. Accordingly, I introduce into a wide-mouthed bottle the tube conducting the gas, and

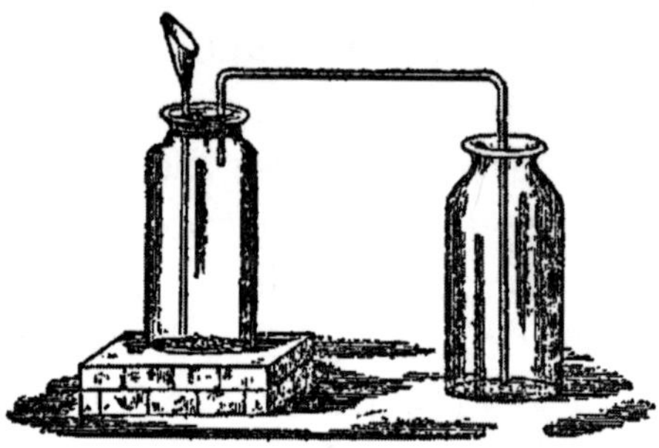

make sure that it reaches to the bottom. In a few moments the bottle will be full of carbonic acid."

"But it will get out of the bottle, for there 's no cork," objected Jules; "or at least it will mix with the air in the bottle."

"You need fear nothing of the sort," replied his uncle. "Carbonic acid is heavier than air. As fast as it arrives, conducted by the tube that reaches almost to the bottom of the receiving bottle, it accumulates and forms an increasingly thick layer, which expels the air, a lighter gas. This air comes out in an invisible stream through the mouth of the bottle, while the carbonic-acid gas takes its place, little by little, from the bottom to the top of the bottle. If we had a vessel full of oil and slowly poured in a stream of water at the bottom, what would happen? The water, being heavier than the oil, would accumulate in the bottom of the vessel, rise gradually, and drive out the oil, the lighter liquid. A similar process goes on when carbonic-acid gas is introduced at the bottom of an enclosed body of air."

"I understand," said Emile, "but I should like to ask one question. With oil, I should know by the color when it was all driven out and its place taken by water; but here there 's nothing to be seen, neither carbonic-acid gas nor air. How can we know, then, when all the air is driven out and the bottle filled with carbonic acid?"

"What our eyes cannot tell us, a flame will help us to see with our understanding. Carbonic-acid is a foe to combustion; it will not keep the smallest flame alive. I will light a piece of paper and thrust it into the mouth of the bottle. If it continues to burn, there is still a layer of air in the upper part of the bottle; but if it goes out, there is nothing in the bottle but carbonic-acid gas. Let us try it; now is the time. Hardly is the lighted paper inside the neck of the bottle when it is extinguished,—a sure proof that carbonic-acid gas reaches up to the mouth of the bottle. We have now a supply of this gas for future experiment. I put the apparatus aside, as we have no further use for it at present. When we need it again, all we shall have to do will be to pour in more hydrochloric acid and let it act on the limestone.

"Here, then, is our carbonic-acid gas. It is as colorless, as transparent, as invisible as air. We have just extracted it from limestone, where chemical combination held great quantities of it captive within a very narrow compass. A piece of stone hardly bigger than a walnut will yield several liters of it. We have just driven some of it out of rock, and we are now going to drive it back and make it reënter the composition of rock,—that is to say limestone, powdered chalk. I pour some lime-water into the bottle that is filled with carbonic-acid gas, close it tightly with the palm of my hand, and shake it thoroughly. The liquid turns white and thick like sour milk. We let it stand a while, and flakes settle at the bottom in a considerable layer. You know these white flakes as carbonate of lime, chalk, the compound we obtained when we shook up lime-water and carbonic acid obtained by burning charcoal. So here we have fresh proof, to add to the others, that limestone really contains the gas produced by burning charcoal.

"The gas has disappeared, being shut up once more in the stone,—or, rather, in a sort of mud that would become stone if it were dried and pressed. Again I bring out our apparatus and renew the supply of carbonic-acid gas in the bottle. How do you suppose a lighted candle would behave in such an atmosphere?"

"It would go out," was Emile's reply, "just as the lighted paper did."

"Besides," added Jules, "nothing can burn except in oxygen or in the air."

The candle did indeed go out, the very instant the flame got as far as the neck of the bottle. The results attained with nitrogen had not been quicker or more complete. Not only was the candle immediately extinguished, but there was not the slightest glow left for an instant in the wick.

"Without making any cruel experiments," Uncle Paul went on, "we can feel sure that this gas, so manifestly unfit for combustion, is equally unfit for maintaining life. An animal would die in it, and very quickly, just as you saw the sparrow die in an atmosphere of nitrogen. Let us now proceed to prove that carbonic-acid gas is heavier than air. I have turned this comparative heaviness to account in collecting the gas without the help of our bowl of water, so that the proof is in reality already before us. Nevertheless I will give you a still more striking demonstration.

"We will take two bottles holding equal amounts and having mouths of equal size. This one, at my right, is full of carbonic-acid gas. I lower a lighted candle into it, and the candle is immediately extinguished. This other bottle, at my left, is full of atmospheric air. I lower a lighted candle into it, and the flame continues to burn. Now, removing the candle, I take the right-hand bottle and gradually invert it, at the same time fitting its mouth to that of the other bottle. In fact, I do exactly as I should do if I were pouring water from the one into the other. I decant the carbonic-acid gas as if it were a liquid. Nothing is seen to pass from the upper to the lower bottle, nor is anything seen to rise from the lower to the upper; and yet the exchange is made, as we shall soon prove. The carbonic acid, the heavier gas, descends and fills the lower bottle, while the atmospheric air, the lighter gas, rises and fills the upper bottle. After waiting a few minutes for the exchange to become complete, I put the two bottles back in their places and again try the lighted candle. It burns in the right-hand bottle; hence this one no longer holds carbonic acid, but air. It goes out in the left-hand bottle, proof that the latter has exchanged its original contents of atmospheric air for the gas that will not maintain combustion. Thus it is made clear that the bottled carbonic acid has gone down and the bottled air has gone up, the two changing places without getting mixed in the process.

"Now listen to this. In various places there is a constant escape of carbonic-acid gas from the ground, especially in the neighborhood of volcanoes. There are springs of carbonic-acid gas as well as springs of water. The most celebrated spring of this sort is that at Pozzuoli, near Naples. It is known as the Dog's Grotto, from the distressing part a dog is made to play for the amusement of the curious. The grotto is hollowed out of the solid rock, and the air in this cavity is earthy, damp, and warm. Bubbles of gas rise here and there in the mud.

"The keeper of this grotto—for there is a keeper who, for money, shows people the repulsive spectacle I am going to tell you about—takes his dog, ties the animal's legs together to prevent its running away, and lays it on the ground in the middle of the grotto, where he himself remains. There is nothing to arouse the slightest suspicion of danger, no foul odor, no lack of clearness in the atmosphere. Besides, is not the dog's master there, showing no sign of fear, standing in the very middle of the cave? Nevertheless there is the dog giving tokens of distress by its groans; it writhes in frightful convulsions, its eyes become dim, its head falls heavily, and it appears to be at death's door. But at this point its master carries it out of the grotto, unties its legs, and lets it breathe the pure air. Little by little the animal revives; it struggles to its feet; still giddy, it looks about in a dull, stupid manner, and then runs off as fast as its legs will carry it, evidently fearing a second ordeal.

"Does the dog act a part taught it by its master? Has it been trained to play dead in the grotto, where its master stands upright beside it without suffering the least inconvenience? No, the dog really comes near dying, as it knows well enough from passing its wretched life in repeating the act several times a day. It knows it so well that it submits to the experiment with a very bad grace. The minute it sees a stranger approaching in the distance it turns morose and surly, growls and threatens to bite. Its master has to hold it in leash when taking it to the grotto, dragging the animal along while the poor beast shows its reluctance by the drooping of its ears and tail. But when the ordeal is over and the stranger gone, it is full of a foolish joy that shows itself in unmistakable fashion. The wretched animal is completely happy at being allowed to come to life again.

"There is nothing about the famous grotto that

cannot be easily explained. Carbonic-acid gas, as I said, comes up out of the ground. This gas is not fit for breathing; after a few breaths an animal dies. Furthermore, it is heavier than air; and so, instead of dispersing evenly throughout the grotto, it remains close to the ground, where it forms a layer about half a meter thick. It reaches only to the knees of the man as he stands in the middle of the grotto, whereas the dog, stretched on the ground, is entirely immersed in it. The master is not compelled to breathe the harmful gas, and so feels no inconvenience from it; the dog breathes nothing else, and in consequence almost dies. But were the dog's master to lie on the ground as does the dog itself, he would share the poor animal's fate.

"The heavier gas is constantly added to and as constantly escapes through the mouth of the grotto, running out in a sort of stream when the air is calm. No one sees this stream, and one passes through it without suspecting it to be there. It makes no murmurous sound and flows over no bed of pebbles, but runs gently and quietly over the grass. Its presence may, however, be detected with the help of a lighted candle: beyond the limits of this invisible stream of gas the candle burns, but as soon as its flame is immersed in the current it goes out as if it were plunged into water. In this way the stream of gas can be traced for some distance from the grotto, while beyond that it is dissipated by air-currents."

"If it were n't so far away," said Jules, when his uncle had finished his story, "I should like to go and see that wonderful grotto; but I should n't ask to have the dog suffer that terrible torture. I should be satisfied with a test made with a lighted candle, first high up in the grotto and then down near the ground, to see whether it would go out when it was lowered."

"If such a test is all you wish for," replied his uncle, "a trip to Pozzuoli is quite unnecessary, for we can reproduce right here the essential conditions found in the Dog's Grotto. A glass jar shall be our grotto, and we will substitute for the carbonic acid rising from the ground the gas our apparatus yields when supplied with limestone and hydrochloric acid. Here is just the jar we need, large enough and with a wide mouth. I insert the long arm of our bent tube and let it reach to the bottom of the jar. Carbonic-acid gas flows through the tube, accumulates at the bottom of the jar, and is retained there by its weight, forming a layer of increasing depth and displacing an equal volume of air. Nothing tells us how deep this layer is at any moment, for the two gases, carbonic acid and air, are both invisible. Nevertheless, from the activity of the effervescence in our apparatus we may give a shrewd guess as to when the jar is about half full of carbonic-acid gas. Then I stop the flow of gas by disconnecting the jar from the apparatus.

"If I am not much mistaken, now is the time to make this disconnection. Having done this, we now see before us an artificial Dog's Grotto,—that is to say, a glass jar filled with carbonic-acid gas below and atmospheric air above. Look through it. It appears to be all alike, the lower layer of carbonic-acid gas being as colorless, as invisible, as the upper layer of air. Our eyes cannot tell us just where the fatal layer ends and the breathable one begins. Though there is a sharp division between the two, no indication of it is to be detected by the eye.

"Slowly I lower a lighted candle into the jar to test its contents. At first it burns very well, and on being lowered farther and farther it still burns; but at last I reach a point where it begins to turn dim. This is the dividing line between the two layers, and if I lower the candle only a little more it immediately goes out, being fully immersed in the carbonic-acid gas. Here, then, we have what Jules was wishing to see,—a reproduction of the Dog's Grotto and the behavior of a lighted candle there. According to the candle's position, high or low, it burns or goes out.

"Now imagine we have in the jar two very different-sized animals, the smaller one completely immersed in the bottom layer, and the larger one with its head in the top layer. The former will perish in a short time, as it breathes a gas incapable of maintaining life, whereas the latter will suffer no discomfort whatever, having a plenty of pure air to breathe. That is the relative position of the dog and the man in the grotto with its lower layer of carbonic acid gas and its upper one of pure air."

# CHAPTER XXII

## DIFFERENT KINDS OF WATER

"IT would be a grave error, my young friends, to regard chemistry as nothing but a succession of experiments for amusing us in our leisure hours. There is, I admit, hardly anything that is more entertaining than to see a ribbon of iron burning brightly in oxygen, or to send up a little balloon inflated with hydrogen and then make it explode with a bang at the touch of fire. If such tricks as these keep the mind wide awake, so much the better, and what we learn through them will be all the more deeply impressed on the memory; but beware of imagining that chemistry ends there. The pursuit of this science is no such paltry affair; it is a very serious business, and has to do with all that concerns us in the material universe. To-day it is to show us why wine, cider, beer, and other fermented drinks foam.

"There are effervescent wines,—that is to say, wines that make the cork pop out of the bottle and are covered with foam when poured into a glass. To obtain such wines, we bottle them before fermentation is finished. Carbonic-acid gas thus continues to form in them; and as its outlet is closed by the cork stopper, it is forced to remain in the liquid, where it accumulates, but never desists from its attempt to escape. It is this gas that makes the cork pop out with a sharp report when the wire or string holding it securely in place is cut; and this it is, too, that sends the wine foaming out of the mouth of the uncorked bottle and makes the froth on the surface of the liquid in the glass, where it causes a slight crackling sound from the continuous bursting of little gas-bubbles. It is this same carbonic-acid gas, again, that gives its foam to cider, which, as you know, is simply fermented apple-juice; and to it is due the froth on beer, a drink made from barley in a sprouting condition."

"Sparkling white wine and cider," observed Jules, "have a sort of sharp taste, but it is n't disagreeable. I don't know anything about beer, for I have never tasted it. Does that sharp flavor come from carbonic-acid gas?"

"Yes. Carbonic acid is a very mild acid, it is true, but still it has the flavor peculiar to all acids, though in a very moderate degree."

"Then we swallow some of this gas when we drink wine or cider or beer, though it is a gas that can kill people if they take a few breaths of it."

"Carbonic acid is dangerous only when taken into the lungs in some quantity. To our fermented drinks it imparts a slightly acid flavor, not disagreeable, and even wholesome, for it aids digestion. You are to understand that such a substance, though capable of killing when breathed, can be quite harmless to the stomach. No one would venture to hold mouth and nostrils under water for any length of time; the result would be death from suffocation or, as we say, drowning. Water is unfit for breathing, it cannot supply the place of air in our lungs; but, for all that, it is the very best of drinks. Carbonic-acid gas is somewhat like that: it can be drunk without danger, and even to advantage when mixed with a beverage, such as wine or cider; but whoever should undertake to breathe it freely would very soon die.

"There is carbonic-acid gas, supplied by nature, in nearly all the water we drink; and it is partly to this gas and its chemical action that we owe the stony material that we take into our stomachs with the water, and whose office it is to contribute to the growth and maintenance of our bones. However clear it may look, the water we habitually drink is hardly ever pure: it contains foreign substances in solution, as is proved by the thin coating of stony matter that gradually accumulates on the inside of carafes and dims the transparency of the glass. This coating is very hard to remove, as it seems to become one with the vessel to which it adheres. Sometimes it is necessary to use strong vinegar to take it off and restore to the carafe its former transparency. This coating is so resistant because it is of stone, real stone, similar in kind to that used by masons for building; in short, it is limestone. Thus the clearest water, water that shows no impurities whatever, nevertheless may contain stone in solution, just as sweetened water contains sugar, though it cannot be seen."

"Then," said Emile, "when we drink a glass of water we drink with it a tiny bit of building-stone. I should never have suspected it."

"It is very fortunate, my young friend, that we do thus drink a little building-stone, as you express it. Our bodies, in order to grow big and strong, need a good supply of stony matter for making bones, which are to us what its timber frame-work is to a building. This material that is so necessary to us is not made by us, but we get it from our food and drink. Water does its part in giving us limestone. If it did not contain this substance in solution, our bones, which are made chiefly of this material, could not develop properly and we should be puny and weak.

"A simple experiment will show us how limestone dissolves in water. Here in this small bottle is a little clear lime-water. Into it, down to the very bottom of the bottle, I thrust the escape tube of our carbonic-acid-gas apparatus. As fast as the gas passes through the tube into the liquid the latter becomes clouded and whitish. We know the reason: the carbonic-acid gas combines with the lime in the water to form carbonate of lime,—limestone, chalk. So far there is nothing new; but let the gas go on discharging itself into the lime-water, and, when it finds no more lime to combine with, it will become absorbed in the water, if not wholly at least in part. Then we shall see the liquid lose its milky cloudiness, gradually turn clear, and finally become as transparent as it was in the beginning.

"Now it is done, the cloudiness has gone, the

white flakes have disappeared. The chalk is no longer to be seen, and the water has turned clear again. Yet we are none the less certain that in this liquid, in which there is nothing to be seen, there is still the carbonate of lime formed a few minutes ago; but it is in solution and therefore invisible. We have learned something new: water that holds carbonic-acid gas will dissolve a small quantity of limestone.

"Something else we are to learn, and then my demonstration will be finished. If this clear water with carbonate of lime in solution were allowed to stand a few days, the carbonic-acid gas would gradually escape, just as the same gas escapes from wine left standing for some time in a glass; and the lime, no longer held in solution by the presence of carbonic acid, would reappear as chalk-dust, and so the liquid would again take on a milky cloudiness. But this return to the milky state can be hastened: we have only to heat the liquid to drive out the carbonic-acid gas, whereupon the chalk again becomes visible and is deposited as a white powder. Thus it is made plain to us, first, that water containing carbonic-acid gas will hold a small amount of limestone in solution, and, secondly, that this dissolved limestone reappears and forms a deposit as soon as the water loses its carbonic-acid gas either by long exposure to the air or by the action of heat.

"Now, carbonic-acid gas escapes from the soil in many places,—as, for example, in the Dog's Grotto with its stream of suffocating gas. Elsewhere the atmosphere always contains some of this gas, if only what comes from the fuel burned in our fire-places and stoves. Rain, in falling through and washing the atmosphere, and springs gushing up from under the ground thus meet with carbonic-acid gas on their way, and some of this they absorb and carry along with them. Afterward, as they flow over the soil, they are likely to become charged with limestone, which occurs in widespread abundance. Such is the origin of the carbonate of lime found in solution in most water. If now the carbonic-acid gas escapes, little by little, from prolonged exposure to the air, the carbonate resumes its stony form and is deposited on whatever objects may be in the water. In this way are formed the calcareous incrustations, or limestone coatings, that line and sometimes stop up our water-mains and the conduit pipes of fountains.

"To be potable—or, in other words, to be fit for drinking—water should contain a little limestone in solution; and, after what I have just told you about the formation of our bones, you will see plainly enough the reason. But when it contains too much, it is hard to digest and oppresses the stomach. The proper proportion is from one to two decigrams, or about a pinch, to a liter of water. Where there is more, the water is what we call hard; or we say it is heavy, because it weighs on the stomach after being drunk.

"Water is sometimes so rich in limestone that it quickly encrusts any object immersed in it. You may have seen springs or brooks that coat with stony matter any blades of grass or tufts of moss they may encounter in their course, forming thus a kind of light rock called tufa. Some of these calcareous springs are quite famous, as, for instance, that of Saint Allyre at Clermont-Ferrand. The water from this celebrated spring falls on a tangle of brush and is thus turned to spray, which is put to service by persons having objects they wish coated with limestone, such as birds' nests, baskets of fruit, bouquets of foliage and flowers. The carbonic-acid gas that helps to hold the limestone in solution escapes, and the spray deposits a coating of stone, so that birds' nests, baskets of fruit, and bouquets, all appear to be petrified. One would almost think a clever sculptor had carved these objects out of marble. Needless to add, such water is not suitable for drinking."

"I should say not," Emile agreed; "the stomach would get a lining of stone, which would not be very easy to digest."

"The water we use for household purposes never contains such an abundance of stony matter, but it often does have enough to cause inconvenience, especially in washing. You must have noticed that water in which linen has been washed with soap is more or less whitish. This whiteness is not due to the soap, for in pure water, such as rain-water, soap dissolves with hardly any effect on the transparency of the liquid, certainly without turning it to a milky whiteness. If ordinary water whitens with soap, it is due wholly to the stony matter in solution. When water whitens a good deal in laundry work and is filled with clots of soap, it is a sure sign that it contains too much mineral matter. Washing is then rendered difficult, and soap dissolves poorly and is wasted in forming flakes without acting on the impurities in what is being washed.

"Water of this sort is bad, too, for certain kinds of cooking, especially for boiling vegetables such as dried peas, beans, lentils, and chick-peas, more particularly the last-named. The stony material in the water impregnates the peas or beans or other vegetables, and you could boil them all day without making them soft. Water of this kind is, of course, as unfit for drinking purposes as for cooking; it overburdens the stomach with its excess of mineral matter.

"Now that we are on the subject, let us finish enumerating the qualities that water should have in order to be good for drinking. It should hold in solution a little air. We will heat some water, and as soon as it begins to get hot we shall see tiny bubbles rising from the bottom. These bubbles are not bubbles of steam, for the temperature is not yet high enough to make steam. They are bubbles of air, air that was held in solution and is now driven out by the heat. Well, this dissolved air is essential to water used for drinking. If it is not present, the water is somewhat disagreeable to the taste and may even provoke nausea. That is why tepid water that has recently cooled down from the boiling-point is not good to drink. The best water, then, is spring water, running water, because its continual motion brings it into contact with and enables it to absorb the greatest possible amount of air. Stagnant water, on the contrary, water that stands still in some ditch, we will say, and is brought but little into contact with the air, is of inferior quality and often positively injurious to the health, especially when decaying vegetable matter is found in it.

"Ordinary water, as I have said, nearly always has a little carbonic-acid gas in solution. I will add that certain springs contain so much that they effervesce and have a slightly acid taste. These are called effervescent mineral springs, and to them belong Selzer, Vichy, and other well-known springs. The water from these is often used for medicinal purposes.

"But enough on the subject of carbonic-acid gas in water. Let us conclude to-day's lesson with a few words on the dangers that lurk in gases composed of carbon combined with oxygen. I say 'gases' and not 'gas,' because the combustion of carbon produces two combinations differing from each other in the amount of oxygen they contain. The one that is burned the more,—or, in other words, oxidized the more, and so is the richer in oxygen,—is carbonic-acid gas, with which we are now well acquainted; the one that is burned less completely, and so is less rich in oxygen, is called carbon monoxid. Unquestionably the first-named is a formidable gas and one to be guarded against, as it is likely to accumulate where it will be a menace to human life, and does so to a notable extent in wine-cellars. Any one forced to breathe such an atmosphere even for a few minutes would surely die unless there were help at hand to resuscitate the victim. Yet carbonic-acid gas is not a poison. We drink a little of it in ordinary water, and much more of it in effervescent drinks; we eat it, so to speak, in our daily bread, which is full of pores made by this gas when the dough fermented; we breathe it constantly, as it is always present in the atmosphere about us; and, finally, the human body itself is a perpetual source of carbonic-acid gas thrown off in breathing. It is plain, then, that it is not a poison. If it causes death on being breathed in an unmixed state, that is due to no injurious properties in the gas itself, but merely because it cannot supply the place of air, the only breathable gas we have any knowledge of. Nitrogen will cause death in the same way.

"Carbon monoxid is quite different: it is really poisonous, a very harmful gas that acts fatally even when breathed in only small quantities and mixed with a good deal of air. It is the more dangerous in that it forms daily in our houses and nothing shows its presence. It is invisible and odorless, an enemy that makes itself known only when the damage has been done. We hear from time to time of some unfortunate person who, either by inadvertence or, as occasionally happens, by intention and from lack of courage to continue the battle of life, dies an untimely death in a closed room containing a charcoal heater. Carbon monoxid is the cause of these lamentable occurrences. Inhaled even in a small quantity, it provokes first a violent headache and general discomfort, then loss of feeling, giddiness, nausea, and extreme weakness.

While this state continues life is in danger, and death may come at any moment.

"It will be well for us to know under what conditions this terrible gas is produced. Since carbon monoxid is carbon less completely burned than when it becomes carbonic-acid gas, it is plain that what ever hinders combustion without stopping it entirely tends to produce this gas. If the draft is poor, if the burning fuel lacks a sufficient supply of air, carbon monoxid is the inevitable product of this imperfect combustion. Remember what takes place when a coal fire is started in a furnace. At first, the greater part of the fuel being cold and the draft of air being sluggish on account of this low temperature, combustion is slow and little tongues of blue flame make their appearance. Later, when the fire is burning briskly, these blue flames are no longer to be seen. Well, these tongues of flame having that beautiful azure color indicate the presence of carbon monoxid, for this gas shows that color in burning completely and changing to carbonic-acid gas. Whenever you see blue flames over a mass of burning coal, you may be sure carbon monoxid is feeding those flames.

"I have now told you enough to make you understand the risk we run when coal or charcoal is burned in such a manner that the products of combustion escape into the room where we are, instead of passing up the chimney; and the risk is all the greater if the room is small and tightly closed. Such a room should never be heated by a brazier, in which combustion is always sluggish, and which always gives out more or less of this fatal gas that does not betray its presence by any sign, but comes upon us treacherously and by surprise. Death may occur even before any danger is suspected. The headache often felt when one is near a stove or a brazier, or even a foot-warmer containing live coals, is the only warning this terrible gas gives. Let us heed this warning and look to our safety.

"It is always very imprudent to close the damper of a bedroom stove in order to keep a low fire during the night. The smoke-pipe being thus closed by the damper and affording no sufficient outlet for the products of combustion, the draft is checked and carbon monoxid forms and spreads through the room, suffocating the sleepers. If a room is small and unventilated, a foot-warmer containing live coals is enough to give a headache and even cause more serious results."

## CHAPTER XXIII

### PLANTS AT WORK

"I SHALL never forget," Uncle Paul resumed, "how rudely one of my friends was treated by a noted cook. One gala day he found the kitchen artist meditating on the triumphs of his profession, as he stood watching the various processes going on in his steaming pots and pans. Broad of face, with a wealth of chins one under another, an opulent nose richly adorned with pimples, a majestic stomach, napkin tucked under his waistband, cap of snowy linen—such was the man.

"The saucepans were simmering on the stove, and from under the lids came whiffs of odors so delicious that one could almost have dined by smelling them. A fat capon stuffed with truffles and a young turkey decorated with slices of bacon were roasting on the spit. At one side a fat thrush redolent of juniper was bestowing a part of its succulence on a slice of buttered toast.

"'Well,' said my friend, after the customary greetings, pointing to one of the saucepans, 'what masterpiece of your art have we here?'

"'Ragout of hare with plovers,' replied the artificer of tempting dishes, beaming with satisfaction and licking his fingers as he lifted the lid. Immediately there was wafted through the room a smell fit to awaken the demon of sensuality in the most abstemious.

"My friend uttered words of high praise, and then continued:

" 'You are clever, as every one admits; but zounds! it 's not so very difficult a matter to cook good dishes when you have good materials to start with, to make the mouth water when you have a fat capon at your disposal, or to create a most appetizing odor with a brace of plovers. The ideal achievement would be to produce a roast or a ragout with no capon, no hare, no bird or beast of any kind. The old recipe, to make a hare pie, first catch your hare, is too exacting. Hares are not running about for every one to catch. It would be more convenient if we could take something else, something common and easy to get, and make our roast or ragout out of that.'

"The cook was at a loss how to reply, so evidently serious were my friend's words.

" 'What!' he exclaimed, 'a real hare ragout without any hare, a real roast capon without any capon? And you mean to say that you could do that?'

" 'No, not I; I am far from clever enough. But I know some one who is clever enough, and compared with whom you and your fellow-artists are nothing but clumsy bunglers.'

"The cook's eyes flashed; the artist's self-esteem was wounded to the quick.

" 'And what, if you please, does this master of masters use? For I suppose he can hardly produce his delectable dishes with nothing at all to work with.'

" 'He uses rather poor materials. Would you like to see them? Here they are, all complete.'

"My friend drew three small vials from his pocket. The cook took one. It contained a black powder, which the culinary artist felt of, tasted, and held to his nose.

" 'It is charcoal,' he declared. 'You are fooling me. Your charcoal capons must be fine eating! Let me see another of your vials. Ha, this is water, or I am much mistaken.'

" 'You are right; it is water.'

" 'Now for your third vial. Why, there 's nothing in it!'

" 'Not so fast; there is something in it—air.'

" 'Air! Your air capons ought not to lie very heavy on the stomach. Are you in earnest?'

" 'Very much in earnest.'

" 'No joking?'

" 'Not the slightest.'

" 'Your artist makes his capons out of charcoal, water, air, and nothing else?'

" 'Yes.'

"The cook's nose turned blue.

" 'With water, charcoal, and air you say he could make this skewer of thrushes?'

" 'Yes, yes.'

"The cook's nose passed from blue to violet.

" 'And with charcoal, air, and water he could make this pasty of fat goose-livers, and this pigeon stew?'

" 'Yes, a hundred thousand times yes!'

"The nose, now attaining its last phase, turned crimson. The bomb burst. The cook decided that the man was a maniac who was making fun of him. Accordingly he took my friend by the shoulders, spun him round, and propelled him through the doorway, throwing his three vials after him. Then the irascible nose gradually returned from crimson to violet, from violet to blue, and from that to its normal tint; but the demonstration to prove that a capon can be made out of charcoal, air, and water was never carried through."

"Of course your friend was only in fun, was n't he, with those three vials?" asked Jules.

"Not at all. His three vials really contained the wherewithal to make the dishes the cook was preparing. Have I not already shown you that charcoal, or carbon, goes to the making of bread, meat, milk, and countless other things we use as food? Remember the slice of bread left toasting too long, and the mutton-chop forgotten on the broiler."

"I see now. Your friend was speaking of the chemical elements. Carbon is one, and that was in the first of the three vials. How about the other two?"

"The second is just as easy to account for. Hold a pane of glass over the smoke rising from a slice of bread when it is just beginning to burn on the stove, and you will soon see the glass covered with a fine film of moisture, exactly as if you had breathed upon it. This moisture comes from the rising vapor, and the latter from the bread. Therefore bread contains water, considerable water in fact, however dry it may appear to be. If we could extract all its moisture from a mouthful of bread, you would be surprised at its quantity. It would astonish you to learn how much water we eat at every meal."

"But we don't eat water, we drink it," objected Emile.

"I say we eat it, for as it is found in bread it does not run, does not wet anything, does not quench thirst. It is solid rather than liquid, dry rather than wet, something to be chewed rather than drunk. Or, better, it is no longer water, but something else that unites with air and carbon to make bread."

"Well," assented the boy, "I 'll admit there must be water in bread because it shows on a piece of glass held over a slice of toast that is beginning to burn. But there is still another vial to account for,—the one with air."

"Here it will be impossible, with the simple means at our disposal, to furnish proof. Of the three substances named as being present in our food I have accounted for two, carbon and water; the presence of the third you must accept on faith."

"Agreed unanimously: there is air in bread. What other wonderful things are you going to tell us about?"

"We will see. This much, first, is agreed upon: bread is composed of carbon, air, and water, which combine in a certain way, merge into one another, and cease to be merely carbon, air, and water, by becoming something else quite different from any one of the three. White has come out of black, savor out of insipidity, nutritive qualities out of innutritious things.

"Meat subjected to the action of fire teaches the same lesson: it turns to carbon and emits fumes containing the constituents of air and water. We will go no farther, for our inquiries would always meet with the same answer. All that we eat or drink, all that goes to nourish us, is reducible to water, carbon, and air. Everything found in an animal's body, everything in a plant, is, with very few exceptions, made of nothing that is not found in water, carbon, or air. Let us be still more explicit. Carbon, being a simple substance, an element, is always carbon and nothing else; but water is composed of hydrogen and oxygen, and air of oxygen and nitrogen. Hence the four elements, carbon, oxygen, hydrogen, and nitrogen, are the materials of which all things in the plant and animal world are almost entirely made.

"So my friend with his three vials was speaking the truth, for all the savory dishes prepared by the cook were reducible to carbon, air, and water. In those little bottles were actually the elements, the prime ingredients, contained in roast capons, pigeon stews, pasties of fat goose-livers, cream tarts, and so on; but to put them together and make them into food, which chemistry in its brutal operations knows only how to destroy, the artist was lacking, the great artist of whom my friend spoke."

"And who is this artist?"

"Vegetation, my young friends, or, more particularly, grass. At the grand banquet spread for all the world, three dishes only are served, though they take an infinite variety of forms. From the epicure who dines on the choicest of delicacies contributed by every quarter of the globe, to the oyster that fills its belly with slime washed up by the waves; from the oak whose roots suck in nourishment from an acre of ground, to the mold on a piece of cheese—all draw upon the same source of supplies, all feed on carbon, air, and water. The only difference lies in the way these ingredients are prepared. Both wolf and man—the latter not unlike a wolf in his kind of food and in other respects also—eat their carbon as it is served to them in sheep, while the sheep finds its carbon in grass, and grass— Ah! here we come to the point that shows vegetation to be the feeder of the world, with wolf, sheep, and man wholly at its mercy.

"In animal flesh both man and wolf find carbon, air, and water served up in compact form as a savory dish, while in grass the sheep finds them just as skilfully prepared, though less savory and less compact in form. But vegetation itself, so nourishing to the sheep, so well fitted to be made into sheep's flesh for the sustenance of man and for the building of his body—with what sauce does it eat its carbon and air and water?

"It eats them with no sauce whatever, but in their natural state, or very nearly so. Blessed with a stomach truly marvelous in its capabilities, the plant digests carbon and takes in air and water, and out of these three substances, which no other form of life will deign to touch as nutriment, makes the forage that hands on to the sheep its needed supply of carbon, oxygen, hydrogen, and nitrogen, all thenceforth united in nutritious form. The sheep takes up the preparation of these elements as found in the blade of grass and carries it further, improving it a little and turning it into flesh that finally, by the slightest possible change, becomes man's or wolf's flesh, according to the consumer."

"I begin to see how it all comes about," said Jules: "man makes his flesh from sheep's flesh and from the various other things he eats, the sheep makes its flesh from the grass it browses, and grass is made of carbon and the elements in air and water. So it is the plant that, in the beginning, prepares all our food for us."

"Yes, the plant, and only the plant, has that important task. Man gets the materials of his body either from the plant itself or from the sheep and other animals that contain those materials already prepared; the sheep or other grazing animal gets them from the plant, where they are found in an advanced stage of preparation; and the plant alone gets them from the original source, eating the uneatable, carbon and the elements in air and water, and, by a marvelous process of which it alone knows the secret, converting them into nutriment fit for the sustenance of animal life. So, then, it is vegetation, finally, that spreads a table for all the earth's inhabitants. Were its work to cease, all forms of animal life, absolutely all, unable to derive nourishment from carbon in its natural state or from air or water, would perish of hunger, the sheep for want of grass, the wolf for want of sheep, and man for want of any and every kind of food.

"I see now," said Emile, "why you called the plant the great artist that knows how to make everything with what was in your friend's vials. It does it all with carbon, air, and water."

"The plant does not eat as we do: it soaks in its food; that is, carbon, for example, is not consumed by it in the natural state known to you as fine black powder, but after it has first been changed from its solid form and dissolved. Now, the solvent of carbon is oxygen, which converts it into carbonic-acid gas, and that is the plant's chief food."

"You say plants live on carbonic-acid gas, that fatal gas that kills us if we breathe a few breaths of it?"

"Yes, my lad, it lives on what would kill us, prepares our food by using what would be death to us if there were enough of it around us. Remember that whatever breathes, whatever burns, whatever ferments, whatever decays, sends out carbonic-acid gas into the atmosphere. The atmosphere, receiving these fatal emanations, would therefore in the course of centuries become unbreathable and suffocating to all animal life on the earth if other agencies did not forestall this accumulation of the deadly gas. Let us now see, to begin with, what statistics have to say as to the amount of carbonic-acid gas that is continually being produced.

"The carbonic-acid gas exhaled by a man in twenty-four hours is estimated at about four hundred and fifty liters, which would represent two hundred and forty grams of burnt carbon and four hundred and fifty liters of oxygen taken from the air to effect this combustion. At this rate the carbonic-acid gas produced in a year by the whole human family would amount to about one hundred and sixty billion cubic meters, which represents eighty-six billion, two hundred and seventy million kilograms of burnt carbon. Piled in one heap, this carbon would make a mountain a league around at its base and from four hundred to five hundred meters high. Such is the quantity of fuel required for the maintenance of man's natural heat. All of us together eat carbon to this extent, and in the course of a year we breathe it out, a breath at a time, in the form of carbonic-acid gas. Then we start on the consumption of another pile of the same size. How many mountains of carbon, then, since the world began, must mankind have breathed out into the atmosphere!

"We have also to take into account the vast variety and extent of animal life on land and in the sea, and its requirements in the way of carbon, requirements that would in the course of a year represent a mountain of perhaps the size of Mont Blanc. Animals far outnumber us, abounding as they do all over the earth, including oceans as well as continents. What a quantity of carbon to keep the flame of life burning! And to think that it all sooner or later goes into the air as a fatal gas, a few breaths of which would quickly kill us!

"Nor is that the whole story. Fermenting substances, like the juice of the grape and like the dough that is to be baked into bread, and all decaying matter,—such as, for instance, is found in the form of manure spread over a cultivated field,—these, too, are rich sources of carbonic-acid gas. Even if the manure is of no great strength it will send out one hundred cubic meters or more of carbonic-acid gas in a single day for every acre of land on which it is spread.

"Coal, wood, charcoal, and other fuel used for heating our houses, and the great quantities of coal needed for running our factories—do not these also contribute to the atmosphere their share of the harmful gas we are speaking of? Just think of the quantity of carbonic-acid gas vomited into the air from the smoke-stack of a great factory that consumes coal by the car-load! Think, too, of the volcanoes, those giant natural chimneys that in a single eruption throw up enough of this gas to make the belchings of a factory furnace seem like a mere whiff on the breeze.

"It is evident that carbonic-acid gas is constantly being poured into the atmosphere in torrents that defy computation; and yet animal life has no reason to fear suffocation, either now or in the future. The atmosphere, continually being tainted, is as continually being purified: as fast as it is laden with carbon it is purged of it. Now, the health-officer charged with the safeguarding of the general physical welfare is the plant, my little friends, the plant that lives on carbonic-acid gas to prevent our dying by breathing it, and with it prepares the food that is to sustain us. This fatal gas in which is taken up so much of the putrefaction of all things is the plant's chief sustenance. To the plant's wonderful stomach, putrefaction is satisfaction. What death has cut down, the blade of grass builds up again.

"It is plain enough that there is never any lack of carbonic-acid gas in the air we breathe, and also that it does not accumulate to the point of becoming a menace to life, as one might at first infer from the abundance in which it is produced. In this dish is some lime-water that I poured out yesterday, when it was perfectly clear. Look at the surface and you will see a delicate transparent crust, which cracks and breaks if you touch it with the point of a pin. It might be taken for a thin sheet of ice. What can it be? The answer is plain: the air, coming into contact with lime-water, has yielded to the latter some of its carbonic-acid gas, and carbonate of lime has consequently formed, not in this instance as a white chalky powder, but as a transparent crystalline sheet."

"I 've often noticed," said Jules, "the same sort of crust on water in which lime has been slaked for making mortar. I should have taken it for ice, but as it did n't melt in the summer sun I concluded it must be something else."

"It was carbonate of lime, formed exactly like that in this dish by the union of carbonic-acid gas from the atmosphere with lime dissolved in the water. Now that our talk has led to this subject, let us say a little more about mason's mortar. You know how it is prepared. The lime-burner first heats a mass of broken limestone very hot in a lime-kiln, and the heat drives out the carbonic-acid gas, leaving the lime by itself. This lime the mason slakes with water and mixes with sand, thus producing mortar, which is laid by the trowel between the stones that go to make a building. At first a soft paste that readily fills in all empty spaces, it gradually becomes impregnated with carbonic acid from the atmosphere, this process being aided by the loosely compacted grains of sand in the mortar, and finally, in consequence of the union of the carbonic acid with the lime, turns to stone as hard and firm as that from which the lime was originally produced. In course of time, therefore, the hardening of the mortar being accomplished, the entire structure of stone and mortar becomes welded into one solid mass, so that often the stones themselves, if one tries to separate them, will break before the mortar will relax its hold. It is, then, carbonic-acid gas from the atmosphere that hardens the mason's mortar by turning the lime it contains back into limestone.

"There is always, as I have said, carbonic-acid gas in the air around us: the hardening of mortar and the forming of a brittle crust on lime-water when exposed to the air prove this sufficiently. But there is not much of this gas thus floating about us; for if the chemist subjects the air to delicate tests such as are beyond the resources of our modest laboratory, he finds that never and nowhere is there more than one liter of carbonic-acid gas in two thousand liters of air. What becomes, then, of the enormous volumes of this gas continually being added to the atmosphere? Vegetation feeds upon it and thus causes it to disappear, as we shall soon see.

"The surface of a leaf is riddled with numerous excessively small holes called stomata, or mouths. On a single leaf more than a million can be counted; but, being so thickly assembled in so small a space, they are of course too small to be seen without a magnifying-glass, a microscope. I cannot show them to you in their natural state, but only as represented in this picture. Well, through these tiny mouths the plant breathes, not pure air as we do, but poisoned air, fatal to the animal, but life-giving to the plant. It inhales, through these millions of myriads of stomata, the carbonic-acid gas dispersed throughout the atmosphere, receiving it into the substance of the leaves, where under the action of the sun's rays an incomprehensible process is accomplished. Stimulated by the light of the sun, the leaves work upon the fatal gas and rid it completely of its carbon. In other words, they decompose the carbonic-acid gas, undo what combustion has done, separate the carbon from the oxygen combined with it.

Stomata, highly magnified.

"You must not think it is any easy task to restore to their original condition two substances combined by burning, by oxidizing. The chemist will have to exercise his utmost ingenuity and use his most potent drugs if he wishes to break the hold that carbon and oxygen have on each other in carbonic-acid gas. Well, it is just this task, which would tax the resources of a chemical laboratory, that the leaves accomplish quietly, without effort, instantaneously, but only on the express condition that the sun gives its aid.

"If the sun does not shine, the plant can do nothing with carbonic-acid gas, which is its chief food. Then it languishes in a half-starved condition, reaching out as if seeking the much-needed light and losing the green color, the hue of health, from its leaves and stems, until finally it perishes. This sickly state caused by lack of light is called etiolation, or bleaching. In market-gardening it is resorted to for the purpose of making certain vegetables more tender and their flavor less strong. Thus it is customary to tie closely together the stems or stalks of a plant used for salad, in this way keeping the sunlight from the central portion and causing it to turn white and tender. So, too, it is usual to heap up the earth around the stalks of artichokes and celery, as their flavor would be unbearable without this treatment of darkness. Lay a tile flat on the grass, or invert a flower-pot over a plant, and after a few days you will find the matted grass or the shaded plant yellow and sickly.

"But when, on the other hand, a plant receives the sun's rays in full force, carbonic-acid gas is decomposed in no time, the carbon and the oxygen separating and each resuming its original attributes. Freed of its load of carbon, the oxygen becomes again what it was before the union took place; it is once more a breathable gas, if diluted with nitrogen, a gas that will support life and make a fire burn. In this pure state it is given out by the stomata to take its part again in combustion and respiration. It entered the leaves a fatal gas, it departs a vivifying gas. It will find its way back some day with a fresh load of carbon, deposit its burden in the plant's storehouse, and, once more purified, recommence its atmospheric round. Bees go and come, one after another, from the hive to the fields and from the fields to the hive, by turn, in the one instance lightened of their heavy load and eager for a fresh burden, and in the other laden with honey and winging their way heavily back to the hive. Oxygen is the swarm of the plant hive: it reaches the stomata with a load of carbon taken from the veins of animals, from burning fuel, or, it may be, from substances undergoing decay, and it gives this carbon to the plant and then departs on its untiring round.

"As to the carbon that the leaves separate from the oxygen, it remains in the plant and enters into the composition of the sap, turning finally into sugar, gum, oil, starch, wood, or some other vegetable substance. Sooner or later these substances are decomposed by the slow combustion of decay, or by the less gradual combustion that takes place in animal nutrition, and the carbon once more enters into the formation of carbonic-acid gas and returns to the atmosphere to feed another generation of plant life, which will hand on to animal life the food products it has with the help of this carbon manufactured.

"Now I will ask Emile if he remembers that old tree stump we were talking about that was some day, it might be, to contribute its carbon to the making of a slice of bread and butter to eat. Does he remember the oak that was to give of its carbon to produce a loaf of white bread or some other article of food? Was I not right in saying it is quite possible that to-day we eat, in the form of a buttered roll, what we some time before burned on the hearth as a stick of wood?"

"I have n't in the least forgotten," replied Emile, "how puzzled I was when you told us those things might some day give us our bread and butter. Now I begin to see how it might come about. A stick of wood burns in the fireplace, and its carbon goes out of it combined with oxygen, making carbonic-acid gas, which is scattered in the air. Plants take in this gas as food and, the next thing we know, the carbon has turned to flour in the grains of wheat, or it may have become grass and been eaten by cows to make butter; and so we have the slice of bread and butter all complete. And then there 's no reason why this carbon from the stick of wood, after it has traveled through the air, should not become a stick of wood again and be burned once more in the fireplace, so that it might go the same round again and again, no one knows how many times."

"Yes, my boy, it can keep going back and forth in that way indefinitely; for the same carbon is forever passing from atmosphere to plant, from plant to animal, and from animal to atmosphere, this last being the common storehouse whence all forms of life derive the principal material of which they are made. Oxygen is the common carrier of this material. The animal gets its carbon from plants or from other animals, in the form of food, and in the end makes carbonic-acid gas out of it, with the help of oxygen, and out it goes into the air. Plants take this unbreathable gas from the air and give back pure oxygen in return, using the carbon in making food for men and animals. Thus the two kingdoms, animal and vegetable, help each other, the former making carbonic-acid gas to feed the latter, and this in turn making breathable air and nutritious substances out of the fatal gas."

"Those are the most wonderful things you have told us yet," declared Jules, deeply impressed by these marvelous transformations. "When you began to tell us about the vials that made the cook's nose turn blue, I thought it was going to be just a funny story, and never dreamed it would turn out to be so interesting and so serious."

"Yes, my child, what I have just told you is indeed interesting and serious, perhaps too serious for one so young as you; but I could not resist the temptation to acquaint you with the beautiful concord existing between the plant that feeds the animal and the animal that feeds the plant.

"Let us now descend from these heights and go on to another experiment. We wish to prove that plants do really decompose carbonic-acid gas. The simplest way to do this is to let the operation go on under water, a method that enables us to observe the release of the gas,—that is the oxygen,—and to collect it. Ordinary water always contains a little carbonic-acid gas in solution, obtained either from the soil or from the atmosphere; therefore we shall not have to supply this to the immersed plant.

"Into a wide-mouthed glass jar filled with ordinary water we put a recently cut plant in full leaf. A water-plant is preferable, as with it quicker and longer-continued results are obtained. We next invert the jar with its mouth in a bowl of water, and then set the whole thing in the sunlight. Soon the leaves are covered with tiny bead-like bubbles, which rise to the upper end of the inverted jar and accumulate there, forming a layer of gas. This gas, as has been proved by trial, will rekindle a match that has just been extinguished, provided it has a spark left; and thus the gas is shown to be oxygen. Hence the carbonic-acid gas dissolved in the water must have been resolved by the leaves into its original elements, the oxygen being set free and the carbon remaining in the leaves.

"But let us put aside our laboratory outfit and resort to the simplest possible proof of the fact in question. Let us go to the nearest pond, where, in some stretch of still water, we shall find a population of tadpoles living and flourishing, either lying in the sun at the water's edge or swimming out into the deeper part, where they frisk and play in freedom. In this pond, also, are to be found various mollusks crawling slowly along, and small shell-fish propelling themselves by jerks as they lash the water with the end of their tail; larvæ, too, encased in little molds of fine sand, and black leeches lying in wait for their prey; and finally, sticklebacks, graceful little fish armed with spines on the back—hence the name.

"All these creatures, of whatever sort, breathe oxygen but it is oxygen held in solution by the water. If the vivifying gas were lacking in the pond, all this teeming population would infallibly perish. Another danger menaces it, also. The bed of the pond is black mud, being an accumlation of decaying matter such as rotting leaves, dead plants, ejections from the aquatic population, the lifeless bodies of various kinds of tiny creatures, and other refuse. This bed of decomposing matter is continually throwing off carbonic-acid gas, as fatal for the stickleback and the tadpole to breathe as for us. How, then, is the water kept clear of this unbreathable gas and enriched with life-giving oxygen for the maintainence of the pond's population?

"Water-plants perform this office of sanitation by feeding on the dissolved carbonic-acid gas, decomposing it under the sun's rays, and giving out oxygen in its place. Decay supports the plant, and the plant supports the animal. Among the various kinds of plants entrusted with the sanitation of stagnant water I will mention the confervæ, that is to say the delicate, green, thread-like growths that overlay the floor of the pond or other body of fresh water with a sort of thick velvet carpet, or float about in jelly-like flakes. Put one of these plants into a bottle of water, and after a short exposure to the direct rays of the sun you will see the plant covered with innumerable tiny beads of gas. These are bubbles of oxygen from decomposed carbonic-acid gas. Caught in the sticky network of the water-plant, these bubbles increase until they finally buoy up the plant and so raise it, all covered with foam, to the surface of the water.

"This experiment calls for no special outfit. Break off a green fragment from some aquatic plant, put it into a glass of water, set the glass in the sun, and soon you will see your little oxygen factory in full operation. When the process is well under way, set the glass in the shade, and the release of gas will stop at once; but put the glass back in the sun, and the formation of gas-bubbles will promptly begin again, thus proving the need of the sun's rays in this wonderful process. I cannot imagine a more beautiful or lucid experiment. It is, too, so simple that I leave you to perform it yourselves.

"The bit of green water-plant making breathable gas in a glass of water set in the sun, will enable you to understand the work of sanitation accomplished in large bodies of water by aquatic vegetation just as a like purification of the atmosphere is effected by terrestrial vegetation. All the green plants of various kinds growing in a standing body of fresh water become covered with tiny beads of oxygen if they receive the sun's rays to help them in their work, and this oxygen dissolves in the water and gives it new life. It is thus by means of the lowest forms of vegetation that standing water, instead of becoming pestilential, may be kept in a condition to support innumerable species of aquatic life.

"From all this you may learn a little lesson that will perhaps be of use to you. How many times you have tried to keep sticklebacks alive in a glass jar! But the attempt has always failed. In water not frequently renewed the little fish soon died, perishing as soon as the small amount of oxygen contained in the water was exhausted by their breathing. Hereafter, if you wish to succeed, put a good-sized clump of confervæ into a jar. Plant and fish will help each other, the plant supplying the stickleback with oxygen, and the stickleback providing carbonic-acid gas for the plant; so both will prosper even in unrenewed water. In short, then, if you would keep your water animals alive, do not forget to give them their indispensable companions, water-plants."

# CHAPTER XXIV

## SULPHUR

"SULPHUR is so well known to you that I need not describe it. It is found chiefly in the neighborhood of volcanoes, where masses of it are unearthed, sometimes in a perfectly pure state, sometimes mixed with soil and stones. In the latter case it has to be freed from impurities.

"You have seen sulphur burning in oxygen with a beautiful blue flame. This combustion produces a gaseous compound of an intensely strong and penetrating smell and provoking a cough when breathed; and we call this gas sulphurous oxid. In ordinary air sulphur burns more slowly, less brightly, but it nevertheless gives the same compound in the end. It is this sulphurous gas that makes us cough when near burning sulphur, even if it be only the small amount of it in a friction match. What use can we make of this disagreeable gas, the merest whiff of which sets us to coughing as if with the whooping-cough? What service can it render us? That shall be the subject of our lesson to-day. But first go out into the garden and pick me some violets and a rose."

These were promptly brought to Uncle Paul, who then placed a little sulphur on a brick, lighted the sulphur, and held over the flame the bunch of violets, having first moistened them slightly. In a few moments the flowers thus subjected to the action of the sulphurous fumes lost their color and turned quite white. The passing from blue to white could be seen very plainly, and it did not fail to call forth Emile's usual expression of surprise.

"Oh, how funny!" cried the boy, watching the process closely as his uncle held the violets in the gas rising from the burning sulphur. "See how white they turn as soon as they are in the smoke. Some are first half-white and half-blue, and then every bit of blue fades away and the whole bunch is pure white, with the flowers looking almost as fresh as before."

"Now let us try the rose," continued his uncle.

Accordingly, the rose was held over the burning sulphur, whereupon its red color faded in the same way and gave place to white, much to the satisfaction of Jules and Emile, who were already planning to repeat for themselves this wonderful experiment in bleaching, so easy to perform with a piece of sulphur and a few flowers.

"So much for that," concluded their uncle, giving the boys the bleached rose and violets to examine at their leisure. "What I have just done, you yourselves can do with countless other flowers, particularly red and blue ones; they will all turn white on being exposed to the sulphurous gas. Henceforth you will know that the pungent vapor of burning sulphur has the peculiarity of destroying certain colors and, consequently, of leaving whiteness in their place.

"This peculiarity is turned to account in many ways, even having its uses in the household. Let us begin with the simplest of these applications. Here is a piece of white cotton cloth, percale. I stain it with the juice of a very ripe cherry. Now what we have to do is to take out the stain. It would be an all but hopeless task to attempt this with soap, but sulphur smoke will do it perfectly and promptly; for as it bleaches flowers so easily it must also take out a spot made by cherry juice, both flowers and cherries being dyed with a vegetable pigment. I moisten the spot slightly and hold it over a small piece of burning sulphur. To direct the fumes more directly upon the desired spot, I cover the sulphur with a small paper funnel, inverting it so that it may serve as a chimney, and I see that the spot is just over the outlet of this funnel. In a moment or two the reddish tint fades and gives place to white, just as did the color of the rose and that of the violets treated in a similar manner. All we have to do now is to rinse the bleached part in pure water. With this precaution the spot will not reappear. That is the way to treat wine stains, which are so hard to remove by washing, and all spots made by preserved fruit with red juice, such as grapes, curcurrants, strawberries, mulberries, blackberries, raspberries, and the like.

"Let us go on to another application even more interesting. After all possible washing, neither silk nor wool, as they come to us in their natural state, will show that perfect whiteness so necessary if they are to reflect in unimpaired brilliancy and purity the dyes with which they are colored. Nor has straw used in the making of hats the desired whiteness; and the same is true of skins for the manufacture of gloves. Well, to whiten wool, silk, straw, and skins —that is, to bleach out their natural yellowish tinge —we treat them as the violets and the roses and the cherry stain have just been treated. They are first slightly dampened and then hung in a tightly closed room in which a few handfuls of sulphur are burned in an earthen bowl. The room becomes filled with the sulphur fumes, and after a day or two the wool and silk and straw come out exquisitely white.

"Sulphur is used for many other purposes, including some that would hardly occur to you. It can even put out a fire. Yes, my little friends, sulphur, which itself burns so readily, will smother flames."

"But that," objected Jules, "is giving the fire more fuel, and one of the most inflammable of fuels, too. I don't understand it."

"You soon will. What does a fire need to keep it burning? Two things, the one as essential as the other,—fuel and air. Imagine a great fire. Is it not true that if we could cut off its supply of air it would speedily go out, and that this would be much better than pouring on water? Is it not true also that if, instead of giving it air, we could substitute a gas unfit for combustion, such as carbonic acid or nitrogen, the fire would not continue to burn another moment? A lighted candle goes out as soon as it is immersed in either of these gases; and so would the best-fed fire in the world if it were suddenly enveloped by a similar atmosphere."

"I see well enough that if we could pour a perfect torrent of nitrogen or carbonic-acid gas on a fire, so as to crowd out all the air and completely surround the burning mass, the flames would be smothered at once; but we could n't do that."

"Not always. In the open air, I admit, it would be hardly practicable; but in a chimney-flue, for example, it is a different matter. There the fire is confined in a narrow passage to which air is admitted through only two openings, and more especially through the lower one. In such circumstances it is not impracticable to make a stifling gas take the place of the air that would otherwise gain admission. Suppose a chimney to catch fire. In order to extinguish the flames in the quickest and simplest manner, recourse is had to sulphur. Any gas unfit for maintaining combustion and itself incombustible would serve; but it is indispensable that this gas be obtainable quickly and in abundance, also at little expense and without the use of any apparatus. Nitrogen and carbonic-acid gas are here out of the question, as it is a difficult, slow, and costly operation to produce them. Sulphurous oxid, however, we can use, as it is instantly procurable if we have a handful of sulphur to throw on the burning coals in the fireplace under the burning chimney. No other gas could be so easily, quickly, and abundantly produced. On the burning brands in the fireplace we throw sulphur without stint, and then shut out the air by stretching a wet cloth across the opening of the fireplace. The sulphur fumes pass up the chimney, driving out the air and so extinguishing the burning soot."

"All the same," said Emile, "it seems queer to put out a fire with sulphur. I should never have thought of such a thing."

"Still another use of this same gas is worth mentioning. It has to do with disease and its cure. We give the name of parasites to all kinds of little creatures that live on other creatures, usually establishing themselves on or in the victim's body. Despite the noble faculties that make him the king of creation, man himself plays his part as victim in this battle between the devourers and the devoured. He has his parasites that live on his substance, just as cherries and walnuts have their peculiar insect pests. Instances without number, alas, make it all too plain that the general rule is applicable to man without respect for his superiority in the scale of creation.

"Besides those powerful and ferocious animals like the lion and the tiger, in whose claws man is as a mouse in the cat's clutches—besides formidable species of that kind, which we can at least openly contend against—we are delivered over to famished hordes that by reason of their very smallness their multitude, and their secure retreat, are able to defy with impunity our efforts at self-protection. First it is the mosquito, armed with a poisoned lancet and sucking our life's blood from our veins without fear of reprisal, and sounding its war song in our ears at night as if in mockery of our impotent wrath. This war-song is the sharp buzzing made by the insect as it approaches us and searches our skin for a suitable spot in which to plunge its lancet."

"How often I have slapped myself," said Emile, "trying to kill those hateful mosquitos when they came and sang their impudent song in my ears in the dark!"

"We have also the louse and the flea, the former attacking our head and the latter assailing the whole body with irritating bites. Like the mosquito, they are after our blood for food; and they get it, too, unless the utmost attention is paid to cleanliness.

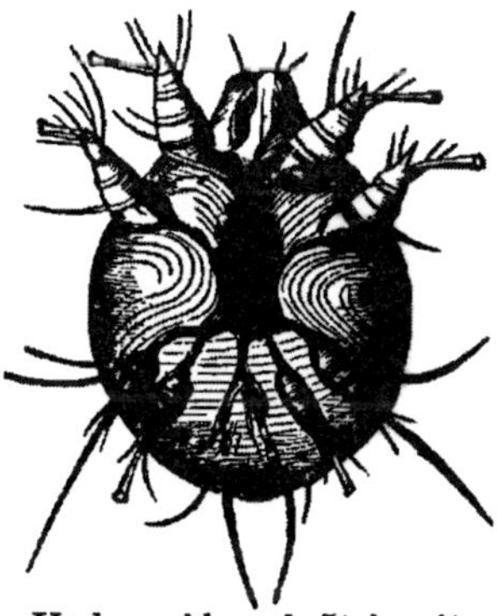

Under side of Itch-mite highly magnified.

"I pass on to another tiny man-eater. A microscopic little creature, an almost invisible parasite called the itch-mite, works its way into our skin, in which it tunnels passages somewhat as a mole burrows a field. Its mole hills are little pimples or pustules that cause acute itching. Such is the origin of the ailment known as the itch."

"You say itch is caused by a parasite that gets into the skin?" asked Jules.

"Yes, and the disease spreads by mere contact, as the parasite passes readily from one person to another, from an infected to an uninfected person."

"And what does it look like, this horrible little creature that makes people scratch so?"

"It looks like a tiny white speck just big enough to be seen by the sharpest eyes, and it is round in shape, reminding one a little of the tortoise. It has eight legs, two pairs in front and two behind, all bristling with sharp, stiff hairs. When it walks it spreads out its eight legs; but in repose it draws them in under its arched body, very much as a tortoise disposes of its legs under its shell. Finally, its mouth is armed with sharp hooks and fine nippers. With these tools it burrows here and there in the skin, making long passages in which it comes and goes at will, just as a mole does in the ground. The learned word for itch-mite means flesh-cutter. I leave you to imagine the unbearable itching this tunneler of human flesh causes when with its well-armed beak it cuts and slashes its way in this direction and in that."

"Don't talk of it any more, Uncle!" pleaded Jules. "It makes me want to scratch just to think of it."

"How is this odious parasite to be got rid of? It is hardly visible and lives inside the skin. Think of trying to catch it there, when it multiplies by the thousand. It is plain that no medicine taken internally will be of any use here. To cure the patient there is but one remedy,—kill the creature that causes the disease. But how kill it in its safe retreat? That is the problem. When a fox commits too many misdeeds and makes havoc among the neighboring poultry-yards, it is smoked out of its hole, stifled in its fortress with burning sulphur until it is forced to surrender. So, too, with the itch-mite. The patient is stripped and enclosed in a kind of tank, his head alone remaining out to allow him to breathe, and then sulphur vapor is made to fill the tank. If this fumigation is properly conducted the insect pest gets enough of it in one turn: it perishes, stifled in its retreat, and the patient is cured. Truly, it is no small service for which he is indebted to the gas we complain of so loudly whenever the fumes of a lighted match tickle our nose.

"To finish our talk on sulphurous oxid, I will tell you about one of its commonest uses. You know how wine is made. I have told you how the sugar in the juice of the grape turns to alcohol by fermenting. Well, this fermentation may go too far, and then the wine sours little by little and turns to vinegar. To prevent this deterioration, the process of fermentation must be arrested; and this is done by burning sulphur in the casks that are to hold the wine. The sulphurous gas thus generated purifies the casks, even working its way into the wood, and stops the fermentation before it goes too far. Thenceforth there is no danger of the wine's turning sour.

"Sulphurous oxid is the only compound that sulphur forms when burned under ordinary conditions, a process you have already witnessed. But I have told you of an acid made from sulphur and containing more oxygen than is found in sulphurous oxid. It is sulphuric acid or oil of vitriol, which we used for decomposing water in order to obtain hydrogen. How, then, is this acid made, this compound of sulphur with an extra supply of oxygen, when we have seen that no matter how much air we give our burning sulphur we get nothing but sulphurous oxid?

"Well, we resort to indirect methods much more powerful than simple combustion. There are, as you know, certain rich storehouses of oxygen,—that is to say, certain compounds in which this gas is found accumulated in unusual abundance, but not firmly fixed. Sometimes a little heat will suffice to set free this captive gas. It was thus that chlorid of potash, heated by means of a few live coals, was made to give us a supply of oxygen. Some of these storehouses bursting with oxygen readily give up a part of their gas to substances that have either none at all or not enough; and to this class belongs nitric acid, or *aqua fortis,* a very useful liquid for oxidizing substances or for adding to the oxygen already in them. If, then, nitric acid is made to act on sulphurous oxid, which is nothing but incompletely burned or oxidized sulphur, the latter takes on additional oxygen from the nitric acid and is changed into sulphuric acid. Immense factories with tall chimneys belching smoke, and with numerous furnaces for burning sulphur, are engaged in the manufacture of this acid, which is used in many branches of industry. Sulphurous oxid, made by burning either pure sulphur or iron pyrites, in which sulphur is abundant, is conducted into enormous leaden chambers shaped like our largest rooms. There it finds nitric acid, which gives up a part of its oxygen, and the transformation is quickly accomplished.

"Sulphuric acid is a heavy liquid, much heavier than water, and oily in appearance, whence its name of oil of vitriol. In a pure state it is colorless, but usually it is impure and of a brownish tinge. Mixed with water, it generates considerable heat. When we were preparing hydrogen, our mixture of sulphuric acid and water became burning hot. The combustion or oxidizing of the zinc, at the expense of the oxygen in the water, had a good deal to do with this generating of heat; but the interaction of sulphuric acid and water had also a part in it. Let us study this latter process by itself.

"Into this glass containing only a little water I carefully pour some sulphuric acid and stir the liquid. The mixture becomes very warm, almost hot. Put your hand on the glass and judge for yourself. What can this heat signify except that a chemical union is taking place between the water and the acid? It shows us that sulphuric acid has a strong tendency to combine with water. The following is another proof of this tendency. Leave a finger's depth of sulphuric acid in a glass for a few days. At the end of that time the liquid will be found to be notably augmented in volume: instead of one finger's depth there may be two. This increase is due to the fact that the sulphuric acid has attracted and incorporated moisture from the surrounding atmosphere. Of course, in the process of gaining volume by taking on water, the acid also becomes weaker. Consequently, if sulphuric acid is to preserve its strength it should be kept in a tightly corked bottle.

"From this affinity for water comes one of the most remarkable properties of sulphuric acid. All animal and vegetable substances are composed chiefly of carbon, hydrogen, and oxygen. Well, if any animal or vegetable substance comes in contact with sulphuric acid, the latter promptly seizes upon the hydrogen and the oxygen to form water, which it appropriates, leaving the carbon by itself exactly as if fire had acted upon the substance. Thus all animal or vegetable material subjected to the action of sulphuric acid is carbonized,—that is to say, reduced to carbon, so that one would think it had passed through fire. Here, for instance, is a pine chip. I immerse it in sulphuric acid and leave it alone a few minutes. You can see the wood turn black; it is reduced to carbon, or charcoal. Fire could not have done it more quickly.

"But I now come to an experiment that will be of even greater interest to you. Into five or six thimblefuls of water I put one drop of sulphuric acid, and no more. It makes a liquid that is unbearably sour like lemon juice, though it looks exactly like pure water. This perfectly colorless liquid I am going to use as writing-ink, very black ink. I take a goose-quill and not a steel pen, as the latter would be attacked by the acid and might discolor it somewhat. My paper is common white paper without any special preparation. Now watch."

Uncle Paul tore a piece of paper out of Jules's copy-book, dipped the goose-quill into the mixture of water and acid, and wrote something in characters that remained as invisible as if he had written in pure water. As soon as all traces of moisture left by the goose-quill had dried up, the paper was handed to the children.

"Read, if you can," said their uncle, "what I have written with my chemical ink."

The paper was carefully examined in the light, first on one side, then on the other, then upside down, and then held against the light; but there was nothing to be seen, the writing being so completely invisible that it was impossible even to guess where the pen had been.

"Your very black ink is not the least black yet," said Emile, "I can't see anything, not the least thing, and if I had not seen you writing I should say this piece of paper had never been used."

"Nevertheless," his uncle assured him, "the invisible is going to become visible. I heat my paper by holding it before the fire. Watch what happens."

At the first touch of heat black characters came out on the white background of the paper as if by magic. Some appeared suddenly and completely, while others came out in separate bits that finally joined one another in unbroken lines, so that very soon there could be read in letters of a deep black: *Carbonization by sulphuric acid.*

"Wonderful!" cried Emile, as he watched the characters forming as if of their own accord. "Wonderful! Let me take your magic ink, Uncle, please; I want to show it to a friend of mine."

"You may keep the magic ink. Weakened as it is by so much water, the acid is no longer dangerous, even in the hands of a giddy-pate like you. Now for the explanation of this writing that so delights you. Paper is made of vegetable matter such as old rags woven of cotton that has first been spun into thread. Hence it contains carbon, hydrogen, and oxygen. Acted upon by the heat from a stove or an open fire, the trifling amount of sulphuric acid in my colorless ink attracts the hydrogen and oxygen at the places touched by the pen, turns these two elements into water, which it appropriates, and leaves the carbon showing plainly in black letters. That is the whole secret. What was at first invisible now shows in marks of a deep black, because the sulphuric acid has laid bare the carbon in the paper.

"What I have just shown you is enough, I think, to make you understand how dangerous this sulphuric acid is, turning everything to charcoal as it does with as much ease as if, instead of being an acid, it were a scorching flame. Whoever handles it should do so with the utmost caution. A single drop on the clothes is first a red spot and then a hole. A drop on the skin is nothing if washed off immediately, but would inflict a very painful wound if left to take effect. It is the eyes, however, that have the most reason to fear this terrible liquid. The least splash of it would lead to the most serious consequences if water were not used at once and in abundance to wash it away.

"And yet this perilous stuff is constantly called into service in a number of industries. Manufacturers find it of the utmost value. Our woven fabrics, our various kinds of leather and glass and soap, our candles, dyestuffs, paper, ink—in fact, a multitude of manufactured articles in common use—all need more or less directly the services of sulphuric acid. I do not mean to say that this acid enters into the composition of a yard of percale, for example, or a sheet of paper, or a cake of soap. What I mean is that its services have been required in the process of making this percale, this paper, and this soap. Sulphuric acid is a necessity in manufacture, an instrument of the most powerful kind, and one that works those transformations without which the manufactured article would never have come into being.

"Take glass as an example. It is made by fusing sand with carbonate of soda. Nature provides the sand all ready for use, but we have to manufacture the carbonate of soda. This is done with the help of sulphate of soda, which itself is obtained by subjecting salt to the action of sulphuric acid. Thus, while glass itself contains no sulphuric acid, this acid is nevertheless needed in the making of glass, as without it salt cannot be made to furnish its soda to the sand that, uniting with this soda, goes to the producing of the glass. A similar part is played by sulphuric acid in the making of soap, which contains a large proportion of soda. Coal for heating our factory furnaces and generating steam for moving the factory machinery, and sulphuric acid for working important chemical changes—these are two of the most potent factors in modern manufacturing industry."

## CHAPTER XXV

### CHLORIN

"WE have more than once had occasion to speak of salt, common cooking-salt and I have told you that it is composed of sodium and chlorin, a metal and a metalloid. In the language of chemistry, salt is chlorid of sodium."

"Are you going to show us some sodium and let us see what it looks like?" asked Emile, his curiosity aroused by the little he had already heard about this metal.

"No, my young friend. Sodium, though not so very rare in drug stores, is too expensive a luxury for our little village laboratory; and so we must content ourselves with a mere description of it. Imagine something that shines like freshly cut lead, and so soft as to yield to the pressure of one's fingers. In fact, it can be molded like wax. Put a piece of it on water and, floating there, it will catch fire and spin around all wrapped in flames. Potassium, the metal contained in ashes, does the same, only more violently. We are now in a position to understand why these two metals have this strange peculiarity of catching fire as soon as they come into contact with water.

"What is water really made of? Oxygen and

hydrogen. Ever since our visit to the blacksmith's shop we have known that red-hot iron decomposes water, appropriating its oxygen and releasing its hydrogen. Unheated iron and also zinc will decompose water in the same way, with no use whatever of fire, if sulphuric acid is added to assist the process. Well, potassium and sodium, as well as some other substances, notably the metal of lime, or calcium, are more active than iron and zinc in the presence of water. Left to play their part unassisted, without heat and without the help of sulphuric acid or anything else, they decompose the water, take to themselves its oxygen, thus becoming oxids, and release the hydrogen. Now, it is this uniting of the oxygen with the metal, this combustion, that develops heat sufficient to set the freed hydrogen on fire; and that explains the flames enwrapping the metal as it spins around on the surface of the water. When the flames die down, the potassium or sodium has disappeared, without leaving the slightest visible trace behind; but the water in which the oxid of the metal was dissolved has now a burning taste and an odor of lye; moreover, it will restore the blue color to litmus previously turned red by an acid.

"If I have no sodium to show you, I can at least let you see the element with which it is joined in partnership in our common salt: I can show you chlorin, which is of more importance than sodium. To obtain chlorin from salt, use is made of sulphuric acid, that powerful agent employed in so many operations in the domain of chemistry. In the present instance the acid's part is to appropriate the metal and thus set the chlorin free. But in order to combine with the acid, the sodium must first be converted into an oxid, into soda; and for this, oxygen is required. The duty of supplying this oxygen is assigned to dioxid of manganese, that same black powder we used in obtaining oxygen from a salt. Its function then was to regulate the heat throughout the mass of salt and thus make easier the decomposition of the latter. Now, however, its purpose is quite different: itself very rich in oxygen, its office is to give up some of this to the sodium, which will thus become soda; and this in turn will then combine with the sulphuric acid to form sulphate of soda. These combinations being effected, the chlorin will be left in a free state, no longer fettered by the bonds hitherto uniting it with the metal sodium.

"The apparatus for this operation is the same as that used in obtaining oxygen. Into our glass balloon I put a handful of common salt and the same amount of dioxide of manganese. I mix them well and sprinkle freely with sulphuric acid. Then, with its tube attached, the balloon is set in place over the brazier so as to receive the heat from the few live coals the brazier contains. A very gentle heat is enough, the release of chlorin beginning even before the temperature has risen at all. Chlorin is a gas heavier than air, and so we will collect it as we collected our carbonic-acid gas; that is, we will arrange the tube from the balloon in such a manner that it will reach to the bottom of our wide-mouthed bottle or jar in which we propose to store the chlorin.

"Up to this point we have had to do only with invisible gases. Air, nitrogen, oxygen, hydrogen, carbonic-acid gas, and carbon monoxid are gases that not even the sharpest eyes can see; and most other gases are of like character in this respect, so that gases as a class are thought of by us as invisible. Now, however, we have a gas that is as subtle and impalpable as the others and yet can be seen very well. It owes its visibility to its pale greenish-yellow tinge. Its name, chlorin, which comes from a Greek word meaning green, takes note of this quality.

"Because of this slight color possessed by chlorin we can watch the gas as it accumulates in the bottom of the jar, where it is held by its weight, and where, also, it displaces the air previously occupying the same space but now driven out by reason of its lesser weight. See there! In the bottom of the jar there appears a kind of fine yellowish vapor, the layer increasing in thickness little by little and filling more and more of the jar from the bottom upward. That yellowish vapor is chlorin; what is above it, colorless and invisible, is air. Wait a few minutes and the visible layer will reach the neck; the jar will then be full of chlorin."

As soon as it was full, the jar was covered with a piece of glass and a second jar filled in the same manner. But during the operation a little of the gas had escaped into the room, perhaps allowed to do so by Uncle Paul in order to teach his pupils how disagreeable chlorin is to breathe. Emile in particular learned the lesson so as never to forget it. Happening to be near the apparatus just when the empty jar was substituted for the full one, he received a whiff of the offensive gas in his nostrils, whereupon he was seized with a fit of coughing such as no cold or whooping-cough had ever caused him. And so our giddy-pate coughed and spat and spat and coughed, but all in vain, in his effort to get rid of what was choking him. It needed his uncle's reassurance to calm his fears as to the consequences of this accident.

"It is nothing, my young friend; you 'll get over your coughing in a few minutes. It was the chlorin that started it, but luckily only a very little of it reached you, and that little was mixed with a good deal of air, so that you smelt the terrible gas more than you swallowed it. Drink a glass of cold water, and that will help to clear your throat."

The cough did in fact soon subside, and the misadventure had no other consequence than to make its victim rather more cautious thereafter about venturing near the jars of chlorin.

"Now you are over-cautious, my child," said his uncle. "A slight smell of chlorin is nothing to be afraid of; it may even be rather wholesome, especially when the air is foul with the products of decay. What is to be feared is breathing this gas pure and admitting it into the lungs in some considerable quantity. Whoever should fill his chest with it, as we fill our chests with atmospheric air, would die in fearful agony after a few breaths."

"I should think as much," declared Emile, "after the fit of coughing I got from just a whiff of it. But how queer it is that common salt should be made of chlorin that suffocates us and sodium that would burn out our mouths if we took the least bite of it! It 's lucky those two fearful substances change so much when they come together, or I should never again dare to salt a radish before eating it."

"It is lucky, too," continued his uncle, "that as soon as it is separated from sodium chlorin regains its energetic properties, for manufacturers profit greatly by these in certain branches of their industry. We will, however, confine ourselves to the one chief use that is made of chlorin, which is in bleaching. Into this jar filled with chlorin I pour some writing-ink from your inkstand there. I shake the jar to make the gas act on the liquid, and in a trice the thing is done. The ink, which was at first of a deep black, has turned to a pale yellow and now looks like slightly muddy water. The liquid so strongly colored at the start is left with hardly any color, the chlorin having destroyed the black of the ink.

"Another way to show the same thing will interest you even more. Here is a sheet of paper written on with ordinary ink, a sheet taken from one of your old copy-books. I moisten it with water to hasten the chemical action of the gas, and put it into our second jar of chlorin. Is it not marvelous to see what now follows? The written characters fade away rapidly, and the paper is left as white as it ever was. I take the paper out of the jar so that you may examine it more closely. Look at it well and tell me whether you can see any trace of the writing that was on it."

The boys gave the sheet of paper the closest

scrutiny, but could not distinguish a single letter. The paper looked as if it had never been used, and the utmost they could make out was an occasional scratch of the pen where it had pressed very hard.

"The writing has all disappeared," declared Jules, "and the paper is as good as new. Would sulphur gas do that? It turns violets and roses white."

"No, it would have no effect. Sulphur is too weak a bleacher. In many cases it is without effect, and here too it would be powerless. Chlorin, on the contrary, has so great a bleaching power that few dyestuffs can withstand it. In this respect it is the most useful agent known to the industrial arts. Nevertheless, not all colors fade away when subjected to the action of chlorin, as a third experiment will prove to you. I take a leaf of an old book of no value and write on it with ordinary ink, even smearing it with blots like a child that does not know how to handle a pen. When the ink is dry, I moisten the page a little and put it into the chlorin. My writing and my ink-blots vanish as by magic, but the print remains as black as ever. The page is now as clean as when first leaving the hands of the printer. The marks I made with writing-ink have been removed, but the printed words have lost nothing of their clearness: they stand out very black against the white background of the paper. The ink-stained and illegible page has been so transformed by the chlorin that it now looks like a leaf from a new book."

"But why is it," asked Jules, "that the printing-ink stays as it was while the writing-ink is all bleached out?"

"That is because the two inks are made of different materials. Printers' ink is made of lampblack and linseed oil. Lampblack is a form of carbon and hence a simple substance, an element, and not something that can be decomposed. Now, chlorin acts on dyestuffs by decomposing them and then combining with one of their elements, hydrogen. Lampblack, which is carbon and nothing more, cannot be decomposed; it cannot give to the chlorin any hydrogen, because it has none to give; hence it remains lampblack and thus keeps its black color. Not so with writing-ink, however: that contains various ingredients, being usually made of sulphate of iron, gallnuts, and logwood. The two latter belong to the vegetable kingdom and contain hydrogen, which the chlorin seizes; and, one of the elements being thus removed, the color disappears.

"The principal use of chlorin as a bleacher is in the manufacture of woven fabrics and of paper. The perfect whiteness of our linen and our writing-paper we owe to chlorin; and so it comes about that before we can have any white paper for writing or printing, or any white cotton or linen for making shirts, handkerchiefs, curtains, and other articles, as well as for making gaily colored calico, we have to call in the services of salt,—or, rather, of that element in salt that we obtain with the help of sulphuric acid and then use as a bleaching agent. There you have an example, to add to the many other examples, of the important part sulphuric acid plays in our manufactures.

"The natural color of hemp and linen is a light reddish, and so fixed is it that it disappears only after repeated washings; hence, the longer a piece of linen has been used and the more times it has passed through the laundry, the whiter and softer it is. To give to linen the utmost possible whiteness, we spread it out on a closely mown meadow and leave it there for weeks at a time, exposed to the sunlight by day and to dampness by night. This prolonged exposure to sun and air, with alternate moistening and drying, finally weakens the hold of the reddish color so that subsequent washings gradually remove it entirely.

"But this method of bleaching is very slow, and when it is applied to great quantities of cloth and over long periods of time, it is also very costly, because it withholds a considerable area of land from productive use. Consequently, in factories where linen and hemp and especially cotton goods are made, a more powerful and less dilatory bleaching agent than sun and dew is called into service; and this agent is chlorin, whose speedy effect on ink you have just seen. Evidently a gas that can take out so black a color as that of ink, and do it so quickly, can easily remove the faint reddish tinge marring the whiteness of hemp, linen, and cotton."

"Wool and silk, too," suggested Jules, "could be bleached with chlorin, and that would be a much quicker way than with burning sulphur."

"But no one except a bungler would for a moment

think of trying it,'' replied his uncle. ''This gas would attack wool and silk and soon reduce them to a mere pulp.''

''But you say cotton, linen, and hemp can stand it.''

''Yes, but their resistance to the action of drugs has no parallel,—a fact that gives them inestimable value. Just think of the many uses we make of linen and cotton and hemp fabrics, and what rough handling they receive,—repeated washing with corrosive lye and the strongest soap, rubbing and beating, and exposure to sun, air, and rain. What sort of material, then, is this that withstands the harsh treatment of washing with soap and of exposure to sun and weather, that remains intact even when all around it is decay, and that defies the manufacturers' powerful drugs and emerges from all these tests whiter and more supple than before? This almost indestructible material is the vegetable fiber of the plants known to us as hemp, flax, and cotton, a fiber unrivaled in its kind. Chlorin, which leaves textiles made of this fiber uninjured, while at the same time bleaching them to an exquisite whiteness, destroys all fabrics made of animal fiber, such as woolen cloth from the sheep's fleece and silk goods from the silkworm's cocoon.

''So common is the use made of chlorin as a bleaching agent that there are many factories devoted solely to its preparation. To make it portable and convenient for use we store it up, as we might say, in lime, which absorbs it freely. This compound is a white powder like lime itself, with a very strong and penetrating odor. It is called chlorid of lime and is a veritable storehouse of chlorin; and it is this that is commonly used whenever a powerful bleaching agent is needed.

"Now I must tell you about the part played by chlorin in paper-making. We commit our thoughts to writing without reflection on the many processes necessary to produce the white paper on which we write. Thousands of years ago the Assyrians of Babylon and Nineveh wrote with a sharp-pointed style on an unhardened clay brick, which was then baked in an oven to fix the writing for all time. If any one wished to send a letter to a friend, it took the form of a brick similar in weight and size to those we use to-day for building."

"With a load of letters like that," said Emile, "postmen to-day would be so weighed down they could n't stagger along."

"If it was desired to write a book," resumed his uncle, "for after ages to read,—a history of the memorable events of the time, for instance,—the work occupied whole shelves of a library, each page of the history being represented by one of these baked bricks. A single one of our printed volumes would have required in the writing enough bricks to build a house. You can judge from that how comparatively small must have been the amount of reading-matter in even a large library of those remote times, when each leaf was so bulky and cumbersome. A few remains of those ancient brick books have come down to us, having been dug up on the sites of Nineveh and Babylon; and these literary remains have even been deciphered and translated.

"Much later, another method of writing, no less strange in our eyes, came into use in those same regions of the East. A reed cut to a pen-point was the writing-instrument, and a black liquid made of soot stirred up in vinegar was the ink, while the paper was a bone, the broad flat shoulder-blade of a sheep, bleached by long exposure to the weather. A packet of writing on one subject—a book, in short—was made up of a number of these bones all tied together with a string.

"In the Europe of long ago, particularly in Greece and Rome, where civilization was most advanced, it was customary to use wooden tablets coated with a thin layer of wax, on which one wrote with a pointed instrument, a style, sharp at one end and having a wide flat head at the other. The pointed end was used for tracing the characters in the wax, the flat end for erasing them and for smoothing the soft surface for fresh writing.

"Of all ancient peoples, the Egyptians came the nearest to inventing something like the paper of modern times. On the banks of the Nile there grows in abundance a kind of reed called papyrus, whose outer covering peels off in long strips, thin and white. These strips were soaked in the muddy water of the river, which served as glue, and were then arranged side by side, with a similar layer over them, but running crosswise. Pressed flat and beaten with a hammer, the whole made a sheet suitable to write on. Here again the pen was a pointed reed and the ink the same liquid prepared from soot. From the word 'papyrus' we get our word 'paper.'

"Papyrus sheets were not cut into small oblong pieces with square corners, such as we are so familiar with, but were made each in one continuous strip, its length varying with the amount of writing to be received. Hence, a papyrus book was all in one sheet or strip, which, for convenience in handling, was rolled around a little wooden cylinder to which the end of it was fastened. When we read a book we turn the leaves one by one, and these leaves have printing on both sides. The ancients followed a different method: they unrolled little by little the long strip of papyrus containing writing on only one side.

"The invention of paper is attributed to the Chinese. In the ninth century the Arabs introduced its manufacture in the nearer East, but its use did not become general in Europe until the thirteenth century. About the year thirteen hundred and forty the first paper factories were established in France. Paper such as you have in the fair white leaves of your copy-books, and such as is used in making our more costly printed books, comes from the despised contents of the ragbag. Shreds and tatters are collected, some being taken out of the mud in the street, and some bearing the marks of unspeakable filth. They are sorted out, the better ones for fine paper, the inferior for coarse. After receiving a vigorous and much-needed washing, they are shredded until the woven fabric is reduced to lint, this process being the work of a cylinder equipped with sharp blades and revolving in a trough containing the rags soaking in water. Thus torn to bits, the rags are at last reduced to a sort of pulp or semi-liquid paste, which is gray in color and has to be thoroughly bleached before it can become the perfectly white paper so familiar to us. This bleaching is done by adding to the pulp, while the cylinder is still in motion, a weak solution of chlorin furnished by the chlorid of lime already mentioned. That is the office of chlorin in paper-manufacture, the bleaching of the rag-pulp to a spotless whiteness.

"But before paper can serve for writing on, it must be prepared in such a way that ink will not soak into it and spread in all directions, thus making the written characters illegible. To this end the pulp or paste receives a certain amount of what is known as size, made of resin and starch. If, however, the paper is intended for use in printing, this preparation is unnecessary; and that is why the paper of our books absorbs ink so freely if we try to write on it.

"The rag-pulp, bleached by chlorin and treated with resin and starch, is now ready for the final operation. With its bits of thread crisscrossing in all directions, it will presently come forth in a thin sheet that will be paper. A machine too complicated to be described here accomplishes this final part of the process. The paste runs in a continuous film over a fine wire netting, which retains the coarser particles and lets the finer pass through. A second and still finer wire netting, moving on rollers, receives what falls from the first, retains the pulp, and drains off the water, the drainage being hastened by a slight side-to-side movement of the netting. In this way the rag-pulp is spread out in a uniform thin layer. Carried onward by the netting on which it is spread, this layer, this undried sheet, is brought into contact with a broad woolen belt, to which it clings, and by which it is conveyed over a hollow cylinder heated within by steam. On this cylinder the paper becomes dry and firm, after which it is rolled up on a second cylinder in a continuous broad strip of indefinite length. A few minutes only are needed to transform the semi-liquid pulp in the trough into paper ready for use. All that has to be done after this is to cut the strip rolled up on the last cylinder into sheets of the desired size.

"In future, whenever you read a printed page or write in your copy-book, remember that we owe the beautiful whiteness of the paper to chlorin, the gas made from our common salt."

# CHAPTER XXVI

## NITROGEN COMPOUNDS

"ON the damp walls of cellars, wine-vaults, and similar places, there is often to be seen a sort of fine white fluff resembling the most delicate down. One might imagine the stone to be covered with a soft fleece. We have already spoken of this curious coating, and Jules has told us how by brushing a damp wall with a feather he has collected some of this material, which, on being thrown on to live coals, has straightway caused a brilliant burst of flame. Its common name is saltpeter, which means salt of stone or salt of rock, because it is on the surface of stones in our buildings or of the rock in caves and underground vaults that this saline matter is found. Chemistry calls it 'nitrate of potash,' a name indicating that it is composed of nitric acid and potash. It is a storehouse of easily obtainable oxygen, which explains why, on being thrown on to live coals, saltpeter makes the coals burst at once into flames. It is the oxygen set free from the saltpeter that produces this result.

"Man is able to make, by going the right way about it—that is, by bringing together the necessary elements—sulphuric acid, sulphurous acid, carbonic acid, phosphoric acid, and many other acids. Burn sulphur, carbon, phosphorus, and there you have, immediately, the last three acids. Sulphuric acid is harder to make, calling for a far more elaborate process than simple combustion. Nevertheless it is made, and in great quantities, too. But nitric acid is quite a different matter: its formation is so hard to bring about that chemists have not yet succeeded in obtaining it by any direct combination of oxygen with nitrogen, the reason for this being the very slight propensity possessed by nitrogen for combining with other elements. It is an inert gas, an inactive element, rebelling against any sort of chemical combination. We have plain proof of this in what takes place every day in our furnaces and stoves and grates. Through the burning fuel, where the temperature is very high, there is constantly flowing a stream of atmospheric air, a mixture of oxygen and nitrogen; yet, despite the great heat, this latter gas does not burn, does not combine with its companion, oxygen, but comes out of the fire the same as when it went in. In short, it is incapable of combustion in the ordinary sense of that word.

"However, what neither chemical skill nor the heat of our furnaces can do, nature accomplishes slowly, noiselessly, without any use of fire, by subtle processes that elude our observation. In the porous substance of a damp stone, the combination of nitrogen with oxygen is brought about, resulting in nitric acid, which finds a little potassium in the stone wall, and so unites with it to make saltpeter. It is to this saltpeter on stone walls and elsewhere that we go for our nitric acid. The method is very simple: all we have to do is to drive the nitric acid out of this compound with a stronger acid. Here the same brutal law prevails as with carbonic-acid gas. 'Get out of here and make room for me,' says the stronger intruder. Sulphuric acid, the indispensable aid in most of these chemical changes, is made to accomplish this displacement. It is added to the saltpeter, and the whole is then heated. Dislodged from its place, the nitric acid escapes as a gas and is collected in a cold receiver, where it condenses as a liquid.

"Now we have obtained nitric acid, so that here we have, chemically united, the two gases (oxygen and nitrogen) that in a merely mechanical mixture constitute the air we breathe. If you were not already well aware of the enormous difference between mere mixture and chemical combination, you would here have a striking example of that difference. The air we breathe and the terrible acid furnished by saltpeter have as constituents the same two elements. I say 'terrible acid,' and the adjective is deserved. Nitric acid is, in fact, so extremely violent in its effects that to denote its potent qualities we often speak of it as *aqua fortis* (strong water). One drop of it on the skin instantly produces a yellow spot, and the result is that the affected skin is burned through and falls off as a dead scale. If the acid is kept in a corked bottle, it speedily corrodes the cork and reduces it to a yellow pulp.

"Metals themselves, even the hardest of them, are eaten by nitric acid. This liquid is a veritable storehouse of oxygen, containing quantities of it and yielding it very freely. Hence, it corrodes or burns most substances that it touches. With its abundant oxygen it causes a sort of combustion, with results that are sometimes the same as those of ordinary combustion. No fire is seen, no flames burst forth, and yet there is really what amounts to combustion, since there is a combination of oxygen with the substance attacked, and this combination is attended by a marked rise in temperature.

"Let us take some examples of this corrosion of metals. I pour a little nitric acid on some iron filings. A dense red vapor immediately rises, while there is a very audible sound and the mixture becomes heated. In a few moments the iron is completely burnt up, turned to rust. I apply the same treatment to this tinfoil, which was wrapped around a cake of chololate, and the same red vapor rises, the same noise is heard, the same increase of temperature is felt. The tin is turned to a white pulp. It is now burnt tin, rusted tin, oxid of tin. I repeat the experiment with copper, and the same results follow, except that the copper-rust is dissolved in the acid as fast as it forms, and produces a greenish-blue liquid. But there are some metals that remain unaffected by nitric acid, and of this number is gold, which never rusts. Here is a piece of gold-leaf such as is used for gilding, and so thin that the slightest breath of air will waft it away. Well, this delicate gold-leaf stays in the acid without showing any effect whatever. It keeps its luster, and will always keep it. Gold does not corrode even if the acid is heated to the boiling-point. I will say in passing that this is the test applied by goldsmiths to distinguish the precious metal from copper, which it so nearly resembles in appearance. Copper is eaten by nitric acid, whereas gold remains unaffected.

"Metal-engravers turn this property of nitric acid to account. When they wish to engrave a copper plate, for example, they first overlay it with an impermeable coating, using melted wax for this purpose. On this coating they then trace the design to be reproduced, and with a fine-pointed instrument remove the wax so as to lay bare the metal wherever they wish it to be eaten away. After this they pour weak nitric acid on the plate thus prepared. Wherever the copper is protected by the wax coating, no effect is produced, but where it is exposed the acid plows a furrow. As soon as the acid is thought to have done its work, the layer of wax is removed and the design is found reproduced in the lines cut into the metal by the corrosive acid.

"So much for nitric acid. Now let us briefly consider its compound called saltpeter or nitrate of potassium. This is used chiefly in the manufacture of gunpowder, which is made by mixing well together, in the right proportions, sulphur, carbon, and saltpeter. So you see there are in gunpowder two highly inflammable substances, sulphur and carbon, together with a third substance, saltpeter, which furnishes an abundant supply of oxygen when it decomposes. Consequently, as soon as gunpowder is set fire to, the saltpeter gives off oxygen freely, and this burns the sulphur and the carbon, which thus become suddenly converted into gas. The amount of gas so generated is enormous. If left free to expand to its full volume, it would occupy one hundred and fifty times the space filled by the gunpowder producing it. Confined, then, to a space much too small for it, this gas makes a vigorous effort to free itself, pushing out of its way with great violence the bullet or ball or anything else that blocks its path, just as a spring forcibly pressed down exerts a powerful thrust on whatever holds it in that position.

"We must now make the acquaintance of another nitrogen compound, one of the utmost importance, especially in agriculture. In this bottle is a liquid that looks exactly like water. I should not, however, advise you to put the open mouth of the bottle to your nose, as your sense of smell would be too painfully affected; but take the slightly moistened cork and give it a cautious sniff. Now what can you tell me about it?"

"*Pfui!*" cried Emile after smelling of the stopper with the utmost circumspection, so suspicious was he of all chemical odors since his experience with the chlorin. "My, how that smarts! It gets up your nose and makes it feel as if it were pricked with a lot of sharp little needles." And he rubbed his eyes, in which the tears were gathering, though he felt not the slightest inclination to cry. He passed the cork to Jules, who at once recognized the liquid by its smell.

"Why, that must be ammonia," he declared.

."It 's what the tailor was using the other day to take out a grease-spot and clean the collar of an old coat. I knew it as soon as I smelt it. Besides, this stuff makes the tears start, and that 's just what the tailor's ammonia did when I got too near it. It was in a cup mixed with water. For a minute or two my eyes were all red and full of tears."

"You are quite right," replied his uncle. "It is indeed ammonia I am showing you in this bottle. It is also called ' volatile alkali ' and ' spirits of hartshorn,' but 'ammonia' is the usual name. A useful property it possesses is that of uniting with grease and making a soluble combination that can be removed by washing. That is why we use it in cleaning garments spotted with grease. With a small stiff brush we first rub diluted ammonia into the soiled places, after which a simple washing with ordinary water will take out the grease. That is what you saw the tailor doing.

"In its composition this cleansing liquid is water containing in solution a large amount of a peculiar gas called ammonia gas. This solution is liquid ammonia or volatile alkali or spirits of hartshorn, its active ingredient being the gas I have just referred to."

"Then liquid ammonia and ammonia gas are two different things?" asked Jules.

"Yes, they are different from each other. Ammonia gas is an invisible, colorless gas that stings the nose smartly and draws tears; but when we speak simply of ammonia we commonly mean the liquid preparation, made by dissolving a great quantity of this same gas in water, to which it imparts its peculiar properties. What I am showing you in this bottle is water in which is stored up an enormous volume of ammonia gas. I say an enormous volume, for in one liter of water there are more than six hundred liters of ammonia gas, the two making together about one liter of liquid. From this well-filled storehouse there is a gradual escape of gas, and that is what has so strong a smell and makes tears come into the eyes. If the gas escaped rapidly, as it would if we heated the liquid, it would overpower us with its pungent odor."

"And it would make us all cry our eyes out, even though we might wish to laugh instead," added Emile. "Chlorin is the gas that makes us cough, ammonia the gas that makes us cry. Each one has a trick of its own."

"That is well said," his uncle agreed. "Ammonia acts strongly on the eyes, making them red and filling them with tears. This peculiarity, together with the pungent odor, enables us easily to detect the presence of this gaseous compound.

"To obtain ammonia gas, we heat to redness certain animal substances of little value, such as old woolen rags, hair, bones, and scraps of leather; and among the gaseous products of the resulting decomposition is the gas we are after. It is collected by simply dissolving it in water. The process of getting illuminating-gas from coal also gives ammonia. The water through which the crude coal gas is passed in order to purify it arrests this other gas in its passage, and finally becomes abundantly charged with it.

"Ammonia gas is composed of nitrogen and hydrogen. On account of nitrogen's disinclination to combine with other elements, any direct union of the two gases so as to produce this compound is as difficult as is the direct production of nitric acid. Chemical science is still unable to make ammonia gas in this direct way, and it is doubtful if it will ever be able to make it thus in any large quantities. This inability is much to be regretted from the farmer's point of view, for ammonia, which to you is merely a good cleaner of soiled clothing, plays a most important part in our fields and gardens, contributing greatly, in the crops it helps to produce, to our daily bread. All forms of life, vegetable as well as animal, contain nitrogen. When they die they give back their elements to the inanimate world by decaying. Their carbon is dispersed in carbonic-acid gas, their hydrogen in water, and their nitrogen in ammonia. But all these products of decay are taken up again by vegetation, the carbonic-acid gas giving carbon, water yielding hydrogen, and ammonia gas supplying nitrogen, while oxygen is everywhere present. Out of these four elements thus assembled by the plant, is built up the substance of our bread, our vegetables, our fruit of all kinds. Refashioned by the animal, which finds it in the plant, this same material becomes flesh, milk, fleece, or some other useful product. In short, nitrogen, in order to reach the animal, must pass through the plant; and in order to reach the plant, the world of lifeless matter must supply it in the combination known as ammonia. We see now why barnyard manure, which is so rich a source of ammonia, is so valuable a fertilizer in agriculture.

"A few words more on ammonia gas dissolved in water: This solution, this liquid ammonia or volatile alkali, is a colorless fluid of the same penetrating odor as the gas itself. It has a burning taste like that of lime and potash; and the resemblance goes even farther, for ammonia has the peculiarity of restoring its original blue color to litmus reddened by an acid. Potash, soda, or lime could not bring back the blue color better or more promptly. We have seen lime turn violets and other blue flowers green. Ammonia, too, turns them green, whether from their natural color or after they have been reddened by an acid.

"The uses of ammonia are numerous. We have spoken of it as a cleanser where grease-spots are to be removed; but it should be added that it will also act on the coloring matter in our garments, changing delicate shades. Hence, it should be used only on material of dark and fast colors that resist the action of this potent cleanser. And here let me tell you something that may be useful to you some day. Those who are engaged in chemical experiments often spill acid on their clothing. Dark-colored cloth is usually turned red by an acid; but a drop of ammonia on the red spot will make it disappear, restoring very nearly the original color.

"Ammonia is also used to counteract the effects

of a venomous sting, such as that of a scorpion, wasp, or bee, or even to prevent the more serious consequences of a viper's bite. Into the little wound a drop of ammonia is poured, and if this is done promptly enough it usually forestalls the action of the venom.

"Finally, ammonia, as it is found in various salts, is a most important food for all plants and vegetables, giving them nitrogen as it does so abundantly. Hence its great value in agriculture; and since manure in the process of decay gives out ammonia gas very freely, it is plain that this dressing must be very beneficial to land under cultivation. But nowadays there is a great demand for artificial fertilizers containing potash and phosphoric acid as well as ammonia."